NONLINEAR SYSTEMS

Modeling and estimation

NONLINEAR SYSTEMS

Modeling and estimation

Edited by

A.J. FOSSARD
ENSAE/CERT, Toulouse, France

D. NORMAND-CYROT
CNRS, Supélec, Gif-sur-Yvette, France

CHAPMAN & HALL

London • Glasgow • Weinheim • New York • Tokyo • Melbourne • Madras

Published by Chapman & Hall, 2-6 Boundary Row, London SEI 8HN, UK

Published with the support of the ministère de l'Enseignement supérieur et de la Recherche (MESR, France): Direction de l'information scientifique et technique et des bibliothèques (DISTB).

Chapman & Hall, 2-6 Boundary Row, London SEI 8HN, UK

Blackie Academic & Professional, Wester Cleddens Road, Bishopbriggs, Glasgow G64 2NZ, UK

Chapman & Hall GmbH, Pappelallee 3, 69469 Weinheim, Germany

Chapman & Hall USA, One Penn Plaza, 41st Floor, New York NY 10119, USA

Chapman & Hall Japan, ITP-Japan, Kyowa Building, 3F, 2-2-1 Hirakawacho, Chiyoda-ku, Tokyo 102, Japan

Chapman & Hall Australia, Thomas Nelson Australia, 102 Dodds Street, South Melbourne, Victoria 3205, Australia

Chapman & Hall India, R. Seshadri, 32 Second Main Road, CIT East, Madras 600 035, India

Co-published by Masson Éditeur, 120 boulevard Saint-Germain, 75006 Paris, France

First English language edition 1995
© 1995 Chapman & Hall and Masson

English language translation revised by Mrs M.B. Groen-Garrer

Original French language edition: *Systèmes non linéaires*, coordonné par A.J. Fossard et D. Normand-Cyrot, © 1993, Masson, Paris.

Typeset in France by Publilog

Printed in France by Corlet SA

ISBN 0 412 59990 2

Preface

Under the aegis of the DRET and the AFCET, the "Nonlinear Group" (NLG) has, over the years, brought together researchers, with their different backgrounds, from specialized schools of engineering, Universities and the CNRS (National Scientific Research Center):

- from the Department of Studies and Research In Automatic Control of the Studies and Research Center of Toulouse associated to the Higher School of Aeronautics and Space ("Département d'Études et de Recherches en Automatique du Centre d'Études et de Recherches de Toulouse associé à l'École Supérieure d'Aéronautique et de l'Espace")
  DERA/CERT/ENSAE

- from the Automatic Control and Computer Science Center of the School of Mining Engineering in Paris ("Centre d'Automatique et d'Informatique de l'École des Mines de Paris")
  CAI/ENSMP

- from the Laboratory of Automatic Control of the State School of Engineering in Nantes ("Laboratoire d'Automatique de l'École Centrale de Nantes")
  LAN/ECN

- from the Laboratory of Automatic Control and System Analysis in Toulouse ("Laboratoire d'Automatique et d'Analyse des Systèmes à Toulouse")
  LAAS/CNRS

- from the Laboratory of Industrial Automatic Control of the INSA in Lyon ("Laboratoire d'Automatique Industrielle de l'INSA de Lyon")
  LAI/INSAL

- from the Laboratory of Automatic Control and Industrial Computer Science of the State School of Engineering in Lille ("Laboratoire d'Automatique et d'Informatique Industrielle de l'École Centrale de Lille")
  LAIL/CNRS

- from the University of Compiègne ("l'Université de Compiègne")
  HEUDIASYC/UTC

- from the Laboratory of Automatic Control in Grenoble ("Laboratoire d'Automatique de Grenoble")
  LAG/INPG/CNRS

- from the Laboratory of Automatic Control and Process Engineering of the University of Lyon I ("Laboratoire d'Automatique et de Génie des Procédés de l'Université Lyon I")
  LAGEP/CNRS

- from the Laboratory of Automatic Control and Industrial Micro-Computer Science of the University of Savoie ("Laboratoire d'Automatique et de Micro-Informatique Industrielle de l'Université de Savoie")
  LAMII/FAST

- from the Laboratory of Signals and Systems in Gif sur Yvette ("Laboratoire des Signaux et Systèmes à Gif sur Yvette")
  LSS/SUPELEC/CNRS

This group, created in 1996 under the initiative of A.J. Fossard, made at the time the following two observations:

• The first one was that substantial progress had been made these last tow years in Nonlinear Automatic Control, in theory as well as in methodology. The number of defended thesis, the papers presented during different conferences, the number of published papers, all widely and publicly revealed the activity of this scientific area. Moreover, this research, which had been in progress for several years, had resulted in a certain number of applications, not only on experimental processes but also on real processes. Even if limited, they clearly translated both a desire for applicability on behalf of the researchers, and a certain maturity of the discipline.

• The second one was that this research effort had not been transferred with desired efficiency at the industrial level, whereas the industrial needs appeared to be more and more obvious, whatever they are related to the slowness of great changes or to the limitations of the controlling organization.

It seemed to us that we, as teachers and researchers, had to assume our responsibilities and should try:

- to have common ideas and a common language,
- to be clear and simple,
- to take into account the real needs, with increased willingness.

Structured by exchanges and reflections, this group therefore wished to make a certain inventory of the knowledge of "modern" automatic nonlinear control, to think about the applicability to new methods on long term and short term, to introduce them in a form which should be as tutorial as possible both in the initial training of engineers and for the necessary transfer to the research departments in industry.

It was largely assisted by the fact that it very soon received material and moral help of the DRET, which had initially been created as a working group of the AFCET. This help appeared to be deterministic in permitting the transformation of what was originally a simple working group into a concerted action, conducted by A.J. FOSSARD and D. NORMAND-CYROT. It is within the framework of this action that the objectives have been specified, both for their substance and form, and that the relationship with the industrial market bag been somewhat institutionalized.

This has not been an easy task. On the one hand because the mathematics which are most often used in nonlinear theory (Lie algebra, differential geometry, etc.) are not taught daily to the engineers, on the other hand because the differences in experience and sensitivity of the members of the group, who in a certain way constitute the richness of it, have created numerous problems for the adaptation to the research of a compromise between rigor and accessibility by the reader.

Each of the different themes treated in the following chapters, has been the object of numerous discussions, drafts, criticisms inside the research group of this difficult compromise. It is this long common work that is presented in the three different volumes, collected under the general theme:

"nonlinear systems"

Each of these volumes is related to a particular aspect:

- Volume 1 Modeling and estimation
- Volume 2 Stability and Stabilization
- Volume 3 Control of nonlinear systems

In this general preface it is out of the question to mention the specific prefaces of each volume, as the title by itself gives already an idea of the problems under study. We will find, however, in the index of each volume, a brief summary of the three volumes together, in order to enable the reader to have a quick view on the contents of the complete work.

Finally, the authors wish to thank all the persons who heave made it possible to work out this document. The DRET morally and materially supported the action of the group. Let us especially mention Mrs FARGEON and Mr ROUCHOUZE whose help and presence have been inestimable. The manufacturers collaborated on this work by their interest that they showed for the group, through their presence at the organized meetings in the framework of this work and through their remarks. The scientific automatic control society supplied with their research the foundations of the presented work, as is shown by the important bibliography at the end of each chapter.

We wish that those books, in accordance with the objectives of the group and thanks to a language which is understood by all of us, will be able to bring together the work of the researchers and the preoccupation of the manufacturers. In that case the objectives will have been completely carried out.

A.J. Fossard and D. Normand-Cyrot
Coordinators of the Concerteu Action of the NG

Contents

Volume 1

Modeling and estimation

Volume 2
Stability and stabilization

Volume 3

Control

Preface

Contents

Volume 3 General introduction

Chapter 1. First order control of nonlinear systems

P. MOUYON

Chapter 2. Input – output linearization

J. DESCUSSE

Contributors

• **Coordinators**

A.J. Fossard
ENSAE/CERT/ONERA
2, av. E. Belin
31055 Toulouse, France

D. Normand-Cyrot
LSS/CNRS/SUPELEC
Plateau de Moulon
91192 Gif-sur-Yvette, France

• **Volume editors**

Volume 1 :
G. Gilles
LAGEP/CNRS
Université Lyon 1
43, bd du 11 novembre 1918
69622 Villeurbanne, France

Volume 2 :
P. Borne
LAIL/CNRS
École Centrale de Lille
BP 48
59651 Villeneuve d'Ascq, France

Volume 3 :
Ph. Mouyon
DERA/CERT/ONERA
2, av. E. Belin
31055 Toulouse, France

• **Contributors**

G. Bornard
LAG/CNRS/INPG
BP 46
38402 St Martin d'Hères, France

C. Burgat
LAAS/CNRS
7, av. du C[l] Roche
31400 Toulouse, France

B. Caron
LAMII/CESALP/FAST
BP 806
74016 Annecy, France

F. Celle-Couenne
LAGEP/CNRS
Université Lyon 1
43, bd du 11 novembre 1918
69622 Villeurbanne, France

A. Charara
Heudiasyc/CNRS/UTC
BP 649
60206 Compiègne, France

G. Dauphin-Tanguy
LAIL/CNRS
École Centrale de Lille
BP 48
59651 Villeneuve d'Ascq, France

J. Descusse
LAN/ECN
1, rue de la Noé
44702 Nantes, France

J. Foisneau
DERA/CERT/ONERA
2, av. E. Belin
31055 Toulouse, France

H.T. Huynh
ONERA
39, av. Division Leclerc
92322 Châtillon, France

J. Lévine
CAS/ENSMP
35, rue Saint-Honoré
77305 Fontainebleau, France

J. Lottin
LAMII/CESALP/FAST
BP 806
74016 Annecy, France

D. Meizel
Heudiasyc/CNRS/UTC
BP 649
60206 Compiègne, France

S. Monaco
Universita di Roma « La Sapienza »
18 via Endossiana
00184 Roma, Italy

L. Pronzato
I3S Sophia Antipolis
250 av. Albert Einstein
06560 Valbonne, France

N.E. Radhy
Faculté des sciences
Aïn Chok, BP 5366, Maârif
Casablanca, Morocco

E. Richard
INRIA Lorraine, antenne de Metz
Technopole Metz 2000, 4 rue Marconi
57070 Metz, France

J.P. Richard
LAIL/CNRS
École Centrale de Lille
BP 48
59651 Villeneuve d'Ascq, France

S. Scavarda
LAI/INSA Lyon
20, rue A. Einstein
69621 Villeurbanne, France

S. Tarbouriech
LAAS/CNRS
7, av. du C[l] Roche
31400 Toulouse, France

D. Thomasset
LAI/INSA Lyon
20, rue A. Einstein
69621 Villeurbanne, France

E. Walter
LSS/CNRS/SUPELEC
Plateau de Moulon
91192 Gif sur Yvette, France

Introduction

Whatever the goal may be (phenomenon understanding, simulation, prediction, control law synthesis, etc...), the knowledge of a dynamic system requires more and more accurate mathematical **modeling**. Searching for this accuracy often leads to a modeling under the form of differential equations (or algebraic-differential equations or partial differential equations) whose structure is nonlinear, either because the nonlinear phenomena play a leading part, or because, under some working conditions, the system has varying dynamic characteristics and it is necessary to take that into account for the stability analysis or for the design of a control law which presents some properties of robustness.

Modeling can be processed by means of a detailed phenomenological analysis and by the application of some fundamental laws from the concerned scientific domains describing these phenomena. We then obtain a **knowledge** mathematical model. In a more pragmatic way of obtaining a control model, we can be satisfied with a **representation** approach, the aim of which is to correctly reproduce the external behavior of the system considered as a "black box". When an adequate structure is chosen for the behavioral model, we must estimate the numerical values of the different parameters involved in this structure. This model design from experimental data is very important in Control Engineering: it is called **identification**. Let us note that parameter estimation is a problem which also occurs in knowledge modeling for the numerical evaluation of unknown parameters involved in the phenomenological equations. Before any parameter estimation is attempted, it is essential to be sure that the collected data includes all the necessary information with respect to the model structure that has been chosen. This is the **structural identifiability** problem which, in the nonlinear world, is analyzed with different mathematical tools according to the classes of systems that are considered.

Another estimation problem arises in Control Engineering, because numerous control strategies - mostly in the nonlinear domain - require the knowledge of the state vector or of a part of it, which is in general not accessible to measurement. The **observation** of a dynamic system consists of real-time estimation of the present state from the input knowledge and the output measurement in a passed time interval. This is only possible if the system model and the collected data satisfy **observability** conditions. These observability and identifiability concepts can be seen from a similar point of view with respect to the model structure and to the richness of information between two similar **estimation** processes: state estimation and parameter estimation. Furthermore when the parameters are time varying, the distinction between observation and identification is not really justified. In the nonlinear case, let us notice that these concepts may present a global or simply local aspect.

It is to this set of concepts, centered around modeling and estimation, that the first volume of this book is devoted, pointing out the nonlinear aspects. The presentation is obviously far from being exhaustive, especially in a domain where, at present, research is very rapidly evolving. We restrict ourselves to the presentation of important concepts, of the main methods and of the current results which seem to be the most directly applicable, and illustrate them with examples. Some of these examples are academic and their goal is to allow the reader to test the understanding of the presented techniques. Others correspond to industrial applications either through simulations or from a true implementation.

At first, modeling is tackled under the phenomenological aspect through approaches that are different with respect to the techniques that are used, but which have in common the analysis of "physical" phenomena that characterize the working of dynamic systems.

The first chapter is devoted to the classical knowledge modeling by means of writing infinitesimal balances of conservative quantities. Through examples, mainly related to the domain of chemical and petrochemical processes, one mentions the variables choice problem and its influence on the numerical solution and the aspects of simplification, robustness and adaptation of the model. The second chapter presents another approach of knowledge modeling: the bond-graph approach. It characterizes the energy exchange phenomena, that is to say the power flux inside a dynamic system, by means of a unique "language", independent from the type of the involved phenomena, taking advantage of the system "architecture" (components or basic elements present, and interconnexions between these components). This language automatically ensures the satisfaction of the principle of energy conservation and power continuity. The characteristic laws of the elementary components can be linear or nonlinear. The computation causality being assigned, such a graphical representation is then very useful in order to build a mathematical model for simulation, pointing out the numerical problems (implicit and algebraic equations, etc.), and also for graphically analyzing the structural properties of the model. In this chapter the different concepts are illustrated by numerous academic examples. Nonlinear applications are finally presented in Power Electronics and especially in Electropneumatics with the detailed phenomenological description of an industrial servojack.

The third chapter deals with the identifiability and discernibility concepts. A given model structure is identifiable if there exists only one value of the parameters associated to a given input-output behavior. A set of model structures is discernible if no parameter values exist, which are associated with models whose structures are different, but which are leading to the same input-output behavior. Without identifiability, the estimation of model parameters from experimental data can become a nonsense, as can the search for the most suitable model structure in the absence of discernibility. Therefore, it is important to test these two properties before collecting and processing the experimental data. In the nonlinear case, these two concepts present a global aspect and a local aspect which are clearly distinguished. The only case which causes difficulties corresponds to the class of systems which are nonlinear with respect to the parameters for which several results are presented. The simpler case of the models whose output is linear with respect to the input, which is a much easier case, is first treated. It enlightens the case where the output is nonlinear with respect to the input, which requires the use of mathematical tools currently utilized in the nonlinear domain: Lie derivation, generating series, nonlinear transformations. The exploitation of these techniques is facilitated by the use of formal computation. Numerous simple examples allow us to verify the understanding of the presented methods and an application in Chemical Engineering is presented in detail.

The fourth chapter deals with nonlinear identification in the case of a particular structure: the discrete time state affine models. In practice, there are numerous time varying continuous linear systems for which we want to obtain a discrete time global model in view of digital control, valid for different operating points. In this case, the state affine models are a good discrete approximation; as a matter of fact, they have a similar role to the continuous bilinear systems. Two approaches are presented for identification. The first one consists in obtaining a bilinear realization from a series of linear identifications, then in building, by means of a regression, a state affine global model. The second one does not require this experimental heaviness: it assumes a single excitation while the influence parameter scans the whole operating range, and an extension of the generalized least square method with the use of a variable parameter linear filter gives the estimation. A software package which is translating the methodology is presented and two examples (helicopter and neutralization chemical pilot plant) illustrate this type of identification/realization.

The last chapter is devoted to the state variable estimation, that is to say, the observation. This is possible only if the system is observable, and the beginning of this chapter naturally deals with the concept of observability. Moreover, if we can define a rank condition which partially generalizes the well-known test of the linear systems, we must notice that, for the nonlinear systems, the observability depends on the inputs. As a matter of fact, some singular inputs can make a nonlinear system unobservable and we have to define inputs which are rich enough in information to provide a good estimation, such as regularly persistent inputs. The nonlinear observation algorithms use the system model and introduce a correcting term proportional (with a variable gain) to the error between the estimated and real outputs, such that the observation error asymptotically tends to zero. The basic idea of the nonlinear observer design consists in trying to progressively extend, under certain conditions, the linear observer synthesis methods, especially the Kalman filter. This is what is done for the state affine systems (including the bilinear case). For the systems that can be split into a state affine part and a nonlinear part satisfying some structural conditions, the techniques of a "high gain" type are applied. Finally, the immersion/output injection techniques allow, via feedback, to globally or locally bring back some nonlinear systems into the class of linear or bilinear systems (with a possible dimension growth) for which we then easily know how to design observers. Finally, some applications are presented, concerning a bioreactor, a mechanical system (overhead crane) and a distillation column.

Gérard Gilles

CHAPTER 1

Physical modeling

G. BORNARD

1.1 Introduction

Elaborating a model is a basic stage in the automation (or the design and optimization) of a process. This ensures a transition between:

- the process, a "real" object, non-formalized, which one can talk about qualitatively,

- the model, a mathematical object, necessary to analysis and synthesis.

The model is expected to enable the user to describe and predict the behavior of the system when the latter is subjected to external influences. One should be able to reason about and to calculate on the model, and to draw conclusions which will be true for the system. That is one of the foundations of the engineer's technical approach, whether the means used is a physical law, an empirical calculation rule, a static model, or a system of differential equations.

Yet, the object and its model are entities of a different nature. Generally, the model is an imperfect and incomplete representation of the reality, whether because of lack of knowledge of certain phenomena, or because of a deliberate simplification satisfying practical requirements. In general, the model is the result of various compromises between the expected results of the model and the means (time, experimental means, computing) necessary to obtain them.

This implies in particular that one does not elaborate "the" model of the studied system, but some satisfactory model which matches the given requirements under certain elaboration and implementation constraints.

1.1.1 Objectives and specifications of modeling

The objectives of modeling affect the specification of the model. We will distinguish three classes of use where the compromise which has to be opted for is different:

- Models for conception:

    - physical representation in a large domain of operation,
    - flexibility of use, easy modification of the structure and of the parameters of the system,
    - usually nonintensive use of the model, few states calculated,
    - tolerance with respect to possible failures of the computing procedure, supervision by the user.

- Models for simulation (test of control systems, education and training):

    - physical representation in a large domain of operation,
    - good computing reliability,
    - intensive use of the same configuration, reduced computing time,
    - more rigid implementation.

- Models for control:

    - absolute numerical computing reliability,
    - intensive use of the same configuration, reduced computing time,
    - model's structure sufficiently simple to enable the design of a control law.

In this chapter, only modeling for simulation and control are considered. Tools for the elaboration of models for conception are described in Chapter 2.

It is clear that economical constraints are also involved in the modeling choices. This aspect – although very important in practice – will not be dealt with in this book.

1.1.2 Categories of models

It is common to distinguish numerous categories of models by combining opposite terms: static – dynamic, continuous – discrete, deterministic – stochastic, with lumped parameters – with distributed parameters, of knowledge – of representation.

Let us develop the last distinction, which characterizes to a large extent the approach adopted to set up a model. Two sources are providing the information which is required to set up a model: the available knowledge about the studied system and the results of experiences on the system itself. Two categories of models are issued from these two sources, to which two approaches of modeling correspond:

- A "knowledge model" (or phenomenological model) is elaborated from a detailed analysis of the system. The behavior of the elements of which is supposed to be governed by rules known from theoretical laws (mechanics, thermodynamics, etc.).

- A "representation model" is elaborated from data resulting from experiments performed on the system.

Representation models are in general relatively simple (linear, bilinear; see Chapter 3). The information included in the measured input and output sequences is the only available information for characterizing the intput-output mapping of such models. As a consequence, they can only realize an interpolation within the studied experimental domain, and extrapolation out of this domain is not valid.

The knowledge models are endowed with predictive properties. In principle, they are able to represent the system in a wider domain. In return, they are in general more complex – in particular nonlinear. Moreover, when applied to a particular process, such models, which are derived from general laws, may significantly differ from the actual behavior of the process.

As a matter of fact, this previous distinction constitutes a methodological point of reference rather than a real model classification in two categories. While numerous regulations presently in operation – linear in particular – have been designed with the help of representation models, the use of pure knowledge models is almost non-existent. Indeed, it is infrequent that all the parameters describing the modelized phenomena are known with an acceptable precision. The behavior of a knowledge model may then be qualitatively correct but requires however an adaptation of the model from experimental data. This is the model's method principle; see for example Richalet *et al.* (1971). The resulting model is recovered from for the two classes mentioned.

Moreover, the complexity of the models directly resulting from the laws of physics often leads to the elaboration of simplified models which include only a part of the initial information already known and which may have to be adapted according to their behavior.

1.1.3   *Contents of the chapter*

This chapter is dedicated to the different problems which arise during the construction of a model. Each stage of the discussion relies on a simple example. These examples have been chosen in order to underline some of the practical difficulties encountered in most of the applications, as well as the methodological references which enable us to solve them.

The approaches adopted in this chapter and in the following one are complementary. Hereafter, the analysis of the diversity of difficulties encountered has been favored. In contrast, Chapter 2 presents a unified and systematic methodological approach to modeling.

All the examples discussed here come from the field of process engineering. The length and unity of the chapter justify this choice. Again in Chapter 2, the stress is put on the electrical and mechanical aspects.

Section 1.2 is an introduction to equation setting and to its interaction with problems of resolution. Section 1.3 is related to the robustness of the model, i.e. its ability to work properly in a large range of situations. The cases of extreme entries and singularity are dealt with. Section 1.4 discusses the simplification of the models, while parametrization and adaptation are studied in section 1.5.

1.2   Equation setting and resolution

The "theoretical" knowledge of a system comes from two sources:

- intrinsic properties, which characterize the constitutive phenomena of the system: heat exchange laws, thermodynamic equilibrium laws, chemical cinetics laws, etc.

- conservation laws of additive species (mass, energy, kinetic moment, etc.) written for subsets of the system, and which characterize its structure.

This knowledge is formalized by setting the equations of the model. This formalization stage reveals information about the system's structure: number of degrees of freedom, size of the state space, etc. It leads to a system of algebraic equations (steady state models) or algebraic-differential equations (dynamic models).

Setting equations implies choices which affect to a large extent the problems appearing in the resolution of the system of equations, i.e. the conditions of use of the model.

Let us consider an elementary example.

1.2.1   Example: single stage separator

The single stage separator under consideration is shown in figure 1.1.

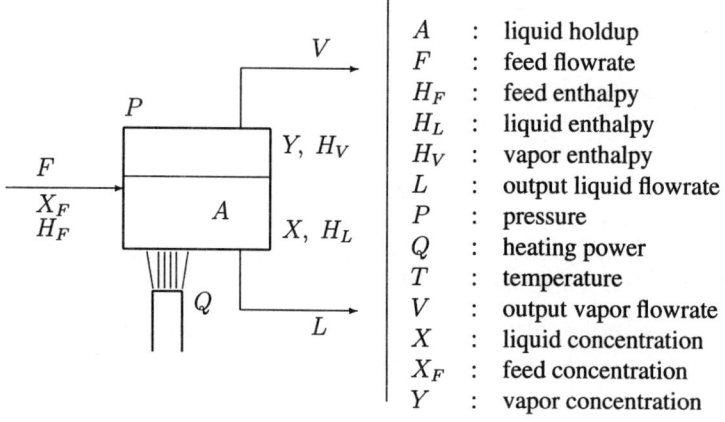

A	:	liquid holdup
F	:	feed flowrate
H_F	:	feed enthalpy
H_L	:	liquid enthalpy
H_V	:	vapor enthalpy
L	:	output liquid flowrate
P	:	pressure
Q	:	heating power
T	:	temperature
V	:	output vapor flowrate
X	:	liquid concentration
X_F	:	feed concentration
Y	:	vapor concentration

Figure 1.1: Single stage separator.

In order to start modeling, let us make the following assumptions:

- the mixture treated is binary (two components);

- each phase is homogeneous, and the phases are in thermodynamic equilibrium (liquid-vapor equilibrium);

- heat losses are neglected;

- the vapor holdup is neglected;

- the pressure is constant (and imposed by the environment).

1.2.2 Steady state modeling

The steady state model is described in order to discuss the link between the use of the model, the choice of the independent variables and the computing algorithm.

1.2.2.1 Equations, variables, degrees of freedom

Two classes of properties enable us to describe the model:

- Intrinsic properties, which characterize the products treated. This is the thermodynamic model of the liquid-vapor equilibrium.

- Conservation laws. These are the material and heat balances.

Since the treated mixture is supposed to be binary, the thermodynamic model may be expressed through the following relations:

$$0 = Y - y(P, X) \tag{1.1}$$
$$0 = T - \tau(P, X) \tag{1.2}$$
$$0 = H_L - hl(T, P, X) \tag{1.3}$$
$$0 = H_V - hv(T, P, Y) \tag{1.4}$$

where y, τ, hl and hv are functions which characterize the treated products. These functions are supposed to have the usual motony properties. The dependence of hl and hv with respect to X and Y, which might be removed in this case, will be useful in section 1.3.1 for the case in which the phase considered is not at the equilibrium temperature.

The conservation laws are expressed through the global material balance:

$$F - L - V = 0, \tag{1.5}$$

the material balance of the light component:

$$FX_F - LX - VY = 0, \tag{1.6}$$

and the heat balance:

$$FH_F - LH_L - VH_V + Q = 0. \tag{1.7}$$

These balances are globally carried out for the whole separator. The process does not require balances on the subsets.

The physically acceptable domain of evolution is characterized by positive flowrates, and concentrations in the interval $[0, 1]$.

Equations (1.1) to (1.7) involve 12 "descriptives" variables: $(F, X_F, H_F, L, V, Q, P, T, X, Y, H_L, H_V)$. The system has 5 degrees of freedom: the number of variables exceeds by 5 the number of equations.

Remark 1

- *The variable A which does not appear in any static equation has however been kept for homogeneity with the dynamic model.*

- *The number of descriptive variables depends on the way the equations are set. For instance the temperature T could have been substituted and disappeared from the equations. Equation (1.2) would have disappeared as well, leaving the number of degrees of freedom invariant.*

1.2.2.2 Independent variables, resolution of the system

Setting $f(x) = 0$ for the equations (1.1) to (1.7), where x is the vector of the descriptive variables, it can easily be checked that the Jacobian matrix $\dfrac{\partial f}{\partial x}(x)$ has a full rank whenever variables L and V are not simultaneously equal to zero. The existence of a nonempty domain of solutions $\mathcal{D}$ will be shown later.

Let $x_0 \in \mathcal{D}$, and consider a partition of x into two disjoint subsets $\bar{x}$ and $\tilde{x}$, where $\bar{x}$ has 5 elements, such that $\dfrac{\partial f}{\partial \tilde{x}}(x_0)$ has a full rank. Then, the implicit function theorem implies that the system has a solution with respect to $\tilde{x}$ for every value of $\bar{x}$ in the neighborhood of $\bar{x}_0$. $\bar{x}$ constitutes a vector of independent variables, and $\tilde{x}$ is the corresponding vector of dependent variables.

The calculation of the dependent variables as functions of the independent variables implies the resolution – generally iterative – of a system of nonlinear algebraic equations.

The various numerical methods and scientific literature which enable us to tackle this problem will not be detailed hereafter. It is, however, interesting to observe, through a few examples, how the choice of the independent variables may affect the complexity of the resulting problem to solve, and what the available possibilities are.

The choice of the independent variables depends on the control or optimization objectives aimed at. Once an independent vector is fixed, it is possible to take directly all the dependent variables as iteration vectors to solve the problem with a suitable numerical method.

A detailed review of the system of equations often put in evidence explicit computing sequences that can be used to reduce the size of the iteration vector before solving it. This approach often leads to better performances and to a more reliable computing, with the counterpart of a previous analysis of the problem. The following situations illustrate these possibilities.

Problem 1: F, X_F, H_F, P are known, and let X_D be a set point for the liquid concentration. Determine the values to be given to Q and L such that the corresponding steady state matches the condition $X = X_D$.

(F, X_F, H_F, P, X) is the independent vector. An explicit resolution is given by:

(*i*)   compute Y, T, H_L, H_V from (1.1) – (1.4),

(*ii*)   compute L, V from (1.5), (1.6),

(iii) compute Q from (1.7),

(iv) end.

Remark 2

- *The presentation of problem 1 refers implicitly to "inputs" and "outputs", which is natural for dynamic systems, but is of no a priori significance for steady-state models.*

- *All the descriptive variables have been computed.*

Problem 2: F, X_F, H_F, P, Q are known, and determine the value of L which leads to a steady state and find the corresponding value of X (reciprocal problem).

(F, X_F, H_F, P, Q) is the new independent vector. The solution is not explicit. It can be solved by the following iteration:

(i) choose a value for X,

(ii) solve problem 1,

(iii) if Q is too far from the desired value, change X (the algorithm is to be specified), and go back to (ii),

(iv) end.

Problem 3: F, X_F, H_F and P are known, and determine the values of Q and L which lead to a steady state for which some function $f(L, V, X, Y)$ reaches its minimum (optimization problem).

An iteration is required, as in the previous case. For instance:

(i) choose a value for Q,

(ii) solve problem 2,

(iii) if the current state is too far from the minimum of f, modify X and go back to (ii),

(iv) end.

A similar procedure can be used by just replacing Q by X, and problem 2 by problem 1. These two procedures are not equivalent: the second is more efficient since the computation made inside the loop is not iterative.

This example illustrates the interaction between the choice of the independent variables and the solving algorithm of the steady-state model. We can notice that, in such cases, there exist opportunities to simplify the resolution.

The problem of the choice of the iteration variables and independent variables according to the situations (simulation, control computing, optimization, estimation of parameters) is analyzed in detail in Bornard (1971).

1.2.3 Dynamic modeling

1.2.3.1 Deriving the equations

The equations (1.1) to (1.4), which characterize the equilibrium remain unchanged. The balance equations now include a term expressing the holdup variations:

$$\dot{A} = F - L - Q \tag{1.8}$$

$$\dot{A}X + A\dot{X} = FX_F - LX - VY \tag{1.9}$$

$$\dot{A}H_L + A\dot{H}_L = FH_F - LH_L - VH_V + Q \tag{1.10}$$

The substitution of (1.8) in (1.9) and (1.10) gives the following equations:

$$A\dot{X} = F(X_F - X) - V(Y - X) \tag{1.11}$$

$$A\dot{H}_L = F(H_F - H_L) - V(H_V - H_L) + Q \tag{1.12}$$

In the system thus obtained, which is composed of the same number of equations and variables as the steady-state model, three algebraic equations are replaced by three ordinary differential equations. In what follows, we assume we know the F, X_F, H_F, P, L, Q, entities which depend on time (and are assumed measurable and bounded).

1.2.3.2 Algebraic-differential system and index

The system obtained is as follows:

$$\dot{x} = f(x, w, u), \quad x \in \mathbb{R}^{n_x}, \quad w \in \mathbb{R}^{n_w}, \quad u \in \mathbb{R}^{n_u} \tag{1.13 - a}$$

$$0 = g(x, w, u) \tag{1.13 - b}$$

where: $x = (A, X, H_L)$, $w = (T, Y, H_V, V)$, $u = (F, X_F, H_F, P, L, Q)$, and where g takes its values in $\mathbb{R}^{n_w}$.

This is an algebro-differential system (ADS), i.e. a system of ordinary differential equations which is coupled with a system of algebraic equations.

A simple case is the case where the Jacobian matrix $\dfrac{\partial g}{\partial w}$ (square) is regular where there exists a function γ such that $g(x, \gamma(x, u), u) = 0$. The ADS is then said to be of index zero.

If γ is explicit, w may be replaced by $\gamma(x, u)$ in (1.13 - a). The system which has to be solved is actually a system of ordinary differential equations and just the fact of having resorted to w when carrying out the model has endowed it with an algebraic-differential form. If g does not have an explicit inverse, its inversion is treated numerically in the numerical integration procedure.

In our example, we have:

$$\frac{\partial g}{\partial w} = \begin{bmatrix} 0 & 1 & 0 & 0 \\ 1 & 0 & 0 & 0 \\ * & 0 & 0 & 0 \\ * & 0 & 1 & 0 \end{bmatrix}$$

Thus, $\dfrac{\partial g}{\partial w}$ is not regular, and the substitution cannot be carried out directly. We are facing an ADS with an index greater than or equal to 1. This notion of index is essential for the numerical integration of an ADS. Efficient numerical integration methods can be found for ADSs with index zero, and, with more difficulty, for ADSs with index 1. For a higher index, it is almost impossible to solve numerically the problem. We will then refer to Hall and Watts (1976), Gear (1971), Petzold and Lötsted (1986).

As Rouchon (1990), showed, the theory of inversion of nonlinear systems permits a clear interpretation of this notion. Let us consider the system:

$$\dot{x} = f(x, w, u), \qquad (1.14\text{-}a)$$

$$y = g(x, w, u). \qquad (1.14\text{-}b)$$

$u(t)$ being known, the resolution of the ADS (1.13 - a, b) corresponds exactly to the following control problem for the system (1.0 - a, b):
The "perturbation u" being known, find the "control law w" such that $y(t) \equiv 0$, which is exactly that of finding the zero dynamics for the system (see vol. III, Chapter 2). The index corresponds to the notion of relative degree for the system (1.0 - a, b).

Our aim is not to detail this aspect but to show how a more careful analysis of the equation setting may help to reduce the difficulty of the numerical problems which are to be solved.

An examination of the model shows that variables A, X and H_L (vector x) are linked by the equation (1.3): the evolution of x takes place in a 2-dimension submanifold.

Substituting (1.3) and (1.9) in (1.12), the latter becomes an algebraic equation:

$$F(H_F - H_L) - V(H_V - H_L) = A\left\{ \frac{\partial hl}{\partial T}\frac{\partial \tau}{\partial X} + \frac{\partial hl}{\partial X}(FX_F - LX - VY) \right\} \qquad (1.15)$$

where P is assumed constant for simplicity.

H_L belongs to w, g contains the new equation (1.15). It is easy to see that the new ADS is of zero index.

Remark 3

We have just described the restriction of system (1.14 a-b) to its output annihilating submanifold (with w as input).

Continuing with this approach, the variables T, Y, H_L, H_V, V can be replaced by their explicit expression resulting from the algebraic equations. The vector w is thus empty, and we obtain a system of ordinary differential equations without algebraic coupling, with A and X as "state variables".

Remark 4

- *It is not always easy or even possible to reduce the relative degree. Rouchon (1990) gives significant examples and a detailed analysis of these matters.*

- *The expressions obtained through a systematic enquiry of the substitutions may be exceedingly complex, and a simpler expression will sometimes be chosen deliberately, at the price of a greater numerical difficulty of resolution.*

1.2.3.3 Discrete time models

If we rely on a "general solver" for differential equations, which is often a good solution, we are generally led to specify a very demanding error bound for integration in order to ensure that coherent results such as material balances are being sufficiently satisfied. This bound is not related to the precision of the whole model, and in many cases generates a considerable amount of computation.

The equations which compose the model must be satisfied with a precision which depends greatly on the properties concerned. The balance equations are mathematically exact, whereas most of the time the thermodynamic laws are subject to a relatively high degree of uncertainty. Besides, very diverse situations may appear in intermediary stages of the computing, such as thresholds on variables, equations with a restricted domain of validity, etc.

Thus, the model developer may seek more directly the control over the computing process so as to improve its reliability, or to manage the compromise between stability, precision and computing time, or even to reach a better physical intuition of the computing process which is used.

The methods for first order implicit integration have good properties, from this point of view. They lead to a direct setting of balances in discrete time which are relatively easy to control. Let us illustrate this point on the separator which we have already studied.

Let us consider the system between time t and time $t + \Delta$, and assume that, for $t < \theta < t + \Delta$:

$$
\begin{aligned}
x(\theta) &= x^\lambda &&= (1 - \lambda)x(t) + \lambda x(t + \Delta) \text{ if } x \text{ is an input, } (P, F, X_F, H_F, L, Q) \\
x(\theta) &= x^+ &&= x(t + \Delta) \text{ otherwise (for stability reasons).}
\end{aligned}
$$

$$(1.16)$$

The value of $\lambda, 0 \le \lambda \le 1$ depends on the determination of the inputs by the surroundings. The equations (1.1) to (1.4) at $t + \Delta$ are given by:

$$0 = Y^+ - y(P^\lambda, X^+) \tag{1.17}$$

$$0 = T^+ - \tau(P^\lambda, X^+) \tag{1.18}$$

$$0 = H_L^+ - hl(T^+, P^\lambda, X^+) \tag{1.19}$$

$$0 = H_V^+ - hv(T^+, P^\lambda, Y^+) \tag{1.20}$$

The discrete time balances are given by:

$$0 = A^+ - A - \Delta\left(F^\lambda - L^\lambda - V^+\right) \tag{1.21}$$

$$0 = A^+X^+ - AX - \Delta\left(F^\lambda X_F^\lambda - L^\lambda X^+ - V^+Y^+\right) \tag{1.22}$$

$$0 = A^+H_L^+ - AH_L - \Delta\left(F^\lambda H_F^\lambda - L^\lambda H_L^+ - V^+H_V^+\right) - Q^\lambda. \tag{1.23}$$

The discrete model appears as a system of algebraic equations. If we choose, for example, X^+ as iteration variable, the solution is obtained with the following iteration:

(i) set X^+ to an initial value,

(ii) find T^+, Y^+, H_L^+, H_V^+ from (1.17) - (1.20),

(iii) find V^+ of (1.23),

(iv) find A^+ of (1.21),

(v) if equation (1.22) is beyond of the fixed tolerance, modify X^+ and go back to (ii),

(vi) end.

This approach may be advantageous for the following reasons:

- The implicit method has good stability properties. A large computing step can be used without leading to a divergence, and time for computing is independent of the step.

- When taking a large step, the quality of the transient response is degraded, but the control may be kept over important phenomena. In the preceding example, this is the case for the material and heat balances, which represent the most certain part of the model and which are satisfied over each sampling period.

- The combination of the two previous points enables a good compromise between precision and computing time.

- The explicit setting of the computing sequence enables the various constraints which are affecting the internal physical variables to be easily taken into account.

In comparison to the use of a general integration method, this approach requires a detailed analysis of the problem to be solved, and thus a longer development time. It is justified when an intensive use of the model is to be made and when the precision of transient responses can be partially relaxed to the benefit of the computing performances (simulation models).

1.3   Robustness of the model

Let us underline the importance of the numerical reliability (robustness) of the models dedicated to simulation or control. Obtaining a robust model implies firstly that the equation setting should generate a problem which is solvable in all the situations that can be encountered. Second, it also implies that reliable numerical methods are used. However, this second aspect, although important, will not be developed here. One shall refer to Brent (1973), Wolfe (1978), Moré and Cosnard (1979) for instance.

The model sought must ensure a sufficiently precise representation of the behavior of the system in a "normal" operating domain. However, the current use of the model may lead it out of this domain under certain circumstances – tests for control, simulation of incidents and failures, etc. This situation, which should be considered in the early stages of the design, is the subject matter of the first section.

The second section considers the case in which the computing reliability may be lost due to singularities in the system of equations to be solved.

1.3.1   Large inputs

According to what has just been said, the model is expected to provide:

- a precise representation in the normal domain of evolution of the process – according to tolerances which are to be specified,

- a "qualitatively correct" and always solvable representation, in all of the domain that can be reached by physically admissible inputs.

Let us assume that the separator model of section 1.2.1 is dedicated to simulation. Under normal operation, the holdup A is maintained near a nominal value by a level regulation which acts on the outflow L for example. However, if the heating power is too high with respect to the supply, or if the control is set in open-loop, the holdup A decreases and may even become null.

The state $A = 0$ is an unusual situation for the process. Qualitatively, the temperature would increase markedly, the power supply would probably decrease because it is produced by a heat exchanger. A precise representation of this situation is useless, and the physical system might already be destroyed.

However, the model must continue to be operational. The model structure has changed – there is no longer a liquid-vapor equilibrium – and the evolution towards $A = 0$ is singular for the equations (1.8) to (1.10).

Qualitatively, the equations (1.1) to (1.4) and (1.8) to (1.10) no longer represent the actual behavior of the system in this region, because of this change of structure and because previously neglected phenomena become preponderant (no liquid-vapor equilibrium, outflow L necessarily equal to zero when $A = 0$, thermal inertia of the iron envelopes, non-homogeneous distribution of the temperatures, etc.).

The discrete model may be extended if, in equations (1.21) to (1.23), we replace L by the variable $\tilde{L}$, defined as follows (this is only an indicative example):

$$\tilde{L}^+ = \begin{cases} 0 & \text{if}\quad A^+ = 0 \\ LA/A_0 & \text{if}\quad 0 < A^+ < A_0 \\ L & \text{if}\quad A^+ \geq A_0 \end{cases} \tag{1.24}$$

$$\begin{cases} A^+ > 0 \implies T^+ - \tau(P, X^+) = 0 \\ T^+ - \tau(P, X^+) \geq 0 \implies A^+ = 0 \end{cases} \tag{1.25}$$

The equation (1.24) forces the output flowrate to be zero when the holdup is zero. The system (1.17) to (1.25) is continuous and has a solution. The discrete computing algorithm is as follows:

(i)   set $A^+ = 0$,

(ii)   find V^+, Y^+ from (1.21), (1.22),

(iii) find X^+ from (1.17), find T^+ from (1.18),

(iv)   find H_V^+ of (1.23),

(v)   if $H_V^+ - hv(T^+, P, Y^+) > 0$: end;

(vi) if not: solve as in 1.2.3.3,

(vii) if $A^+ > A_0$: end;

(viii)if not: solve as in 1.2.3.3 by replacing L with $\tilde{L}$,

(ix) end.

Remark 5

- *The computation of X^+ and H_L^+ could be bypassed when $A^+ = 0$.*

- *The extension of the domain required extra information: the dependence of the flowrate $\tilde{L}$ with respect to A, and the disequilibrium vapor enthalpy.*

- *This example illustrates also the flexibility introduced by the direct setting of a discrete time model.*

1.3.2   Singularities: example of a countercurrent heat exchanger

Let us now examine the case where a singularity appears within the normal operating domain. It will be a "soft" singularity since, as we shall see, the system is continuous near the singularity after division by a small term, ε.

1.3.2.1   Process and assumptions

The process under study – a countercurrent heat exchanger – is described in figure 1.2.

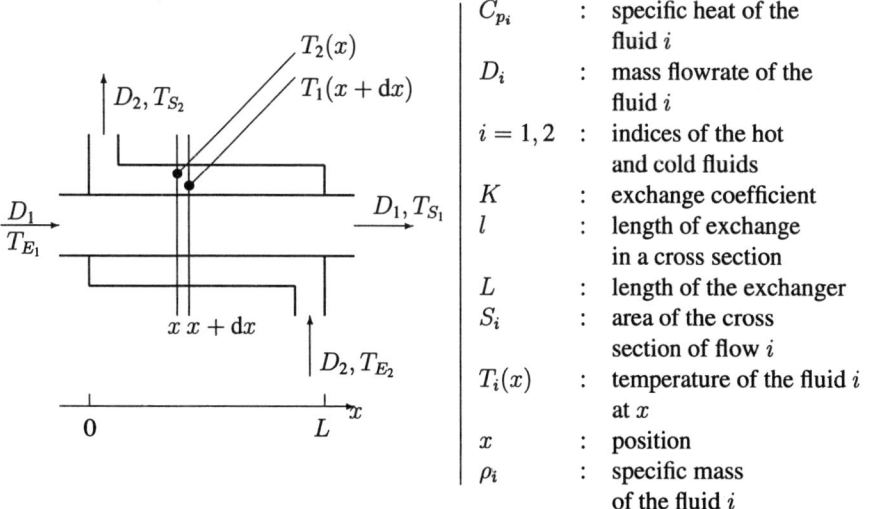

C_{p_i}	:	specific heat of the fluid i
D_i	:	mass flowrate of the fluid i
$i = 1, 2$	:	indices of the hot and cold fluids
K	:	exchange coefficient
l	:	length of exchange in a cross section
L	:	length of the exchanger
S_i	:	area of the cross section of flow i
$T_i(x)$	:	temperature of the fluid i at x
x	:	position
ρ_i	:	specific mass of the fluid i

Figure 1.2: Countercurrent heat exchanger.

In order to derive the equations of the model, the following assumptions are made:

- the fluids are incompressible

- the densities and specific heat are constant

- the exchange coefficient is constant

- the temperature within each cross-section is homogeneous

- the axial diffusion is neglected

- heat losses are neglected

These assumptions are questionable – in particular the constant exchange coefficient. Nevertheless, they are sufficient for the discussion introduced below.

1.3.2.2   Steady state model

This steady state model is intended to be used either directly, or as an intermediate stage in the dynamic model elaboration, see Jaouahri (1980).

The heat balance of the slice $(x, x + \mathrm{d}x)$, in association with the exchange law, leads to the following equations:

- balance of the slice for flow 1:

$$D_1 C_{p_1} T_1(x) - D_1 C_{p_1} T_1(x + \mathrm{d}x)$$

- balance of the slice for flow 2:

$$D_2 C_{p_2} T_2(x + \mathrm{d}x) - D_2 C_{p_2} T_2(x)$$

- exchange from flow 1 to flow 2:

$$K l \mathrm{d}x (T_1(x) - T_2(x))$$

If $\mathrm{d}x \to 0$, one obtains:

$$D_1 C_{p_1} \frac{\mathrm{d}T_1}{\mathrm{d}x}(x) + K l (T_1(x) - T_2(x)) = 0 \tag{1.26}$$

$$D_2 C_{p_2} \frac{\mathrm{d}T_2}{\mathrm{d}x}(x) + K l (T_1(x) - T_2(x)) = 0 \tag{1.27}$$

The input variables are D_1, D_2, $T_{E_1} = T_1(0)$, $T_{E_2} = T_1(L)$. While integrating equations (1.26) and (1.27); two situations appear:

Case 1: $C_{p_1} D_1 \neq C_{p_2} D_2$

    We have:

$$T_1(x) = a + b h_1 \exp\left((h_2 - h_1)x\right) \tag{1.28}$$

$$T_2(x) = a + b h_2 \exp\left((h_2 - h_1)x\right) \tag{1.29}$$

with:

$$\begin{cases} h_1 & = & \dfrac{Kl}{C_{p_1} D_1} \\[2mm] h_2 & = & \dfrac{Kl}{C_{p_2} D_2} \end{cases} \qquad \begin{aligned} b & = & \dfrac{T_{E_2} - T_{E_1}}{h_2 \exp\left((h_2 - h_1)L\right) - h_1} \\[2mm] a & = & T_{E_1} - b h_1 \end{aligned} \tag{1.30}$$

Case 2: $C_{p_1} D_1 = C_{p_2} D_2$

    We obtain:

$$T_1(x) = cx + d_1 \tag{1.31}$$

$$T_2(x) = cx + d_2 \tag{1.32}$$

with:

$$\begin{cases} c & = & \dfrac{h_1 \left(T_{E_2} - T_{E_1}\right)}{1 + h_1 L} \\[2mm] d_1 & = & T_{E_1} \end{cases} \qquad d_2 = \dfrac{T_{E_2} + h_1 L T_{E_1}}{1 + h_1 L} \tag{1.33}$$

1.3.2.3 Behavior near the singularity

In the previous equations, a singularity appears, in the submanifold associated with case 2. Let us consider case 1 where $C_{p_1} D_1 - C_{p_2} D_2 \to 0$:

$$\begin{aligned} h_1 & \to & h_2, \\ |a| & \to & \infty \\ |b| & \to & \infty \end{aligned}$$

$T_1(x)$ and $T_2(x)$ issue from the difference of terms whose absolute values increase arbitrarily. This implies that the truncation error is also increasing so that the numerical result has no significance.

To avoid this difficulty, one can give, in neighborhood of the singularity, a continuous approximation of the solution. Let $\varepsilon = h_2 - h_1$. A limited development of (1.28) and (1.29) as a function of ε gives, after simplification by ε, the following expressions:

$$T_1(x) = f + g h_1 \left(x + \varepsilon \frac{x^2}{2} + \varepsilon^2 \frac{x^3}{6} + \ldots\right) \tag{1.34}$$

$$T_2(x) = f + g \left[h_1 \left(x + \varepsilon \frac{x^2}{2} + \varepsilon^2 \frac{x^3}{6} + \ldots\right) + \exp(\varepsilon x)\right] \tag{1.35}$$

with:

$$f = T_{E_1} \qquad g = \frac{T_{E_2} - T_{E_1}}{1 + h_1 L + (L + h_1 \frac{L^2}{2})\varepsilon + (\frac{L^2}{2} + h_1 \frac{L^3}{6})\varepsilon^2 + \ldots} \qquad (1.36)$$

Figure 1.3 shows the evolution of the errors:

- the truncation error due to the resolution of (1.28) and (1.29) increases as $\varepsilon \to 0$,

- the approximation error linked to the limited development (1.34) and (1.35) is zero for $\varepsilon = 0$ and increases with $\|\varepsilon\|$.

The upper bound of the error can the be managed by an adequate choice of the order of the development and the size of the domain on which the approximated computation is made.

Remark 6

Equations (1.28) and (1.29) could have been organized differently without eliminating the problem set by this singularity.

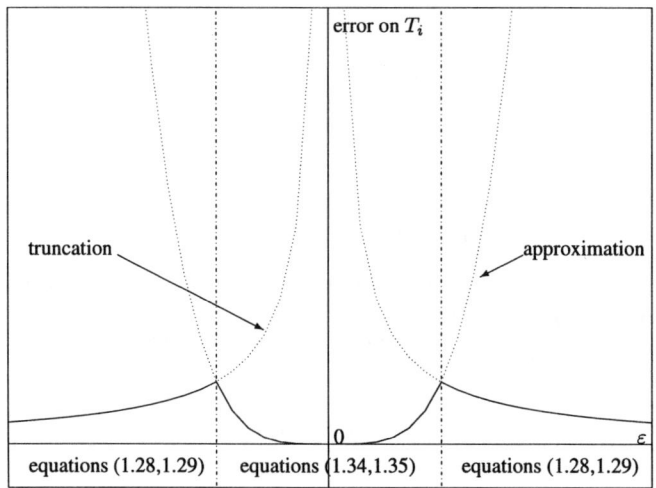

Figure 1.3: Exchanger: truncation and approximation errors.

1.4 Simplification

Equations derived from physics are often complex and commonly lead to a large number of descriptive variables. Simplification of the model may be necessary in order to achieve acceptable complexity, dimension and computing time.

We shall now emphasize the concept of an *a posteriori* simplification. Such a procedure is based on the previous computation of several operating points from a complete basic model. Distillation will be taken as an example illustrating several simplification procedures.

1.4.1 Example: multicomponent distillation column

The scheme given in figure 1.4 represents a conventional colomn treating a multicomponent mixture. However the methology introduced would also apply to columns with a more complex structure.

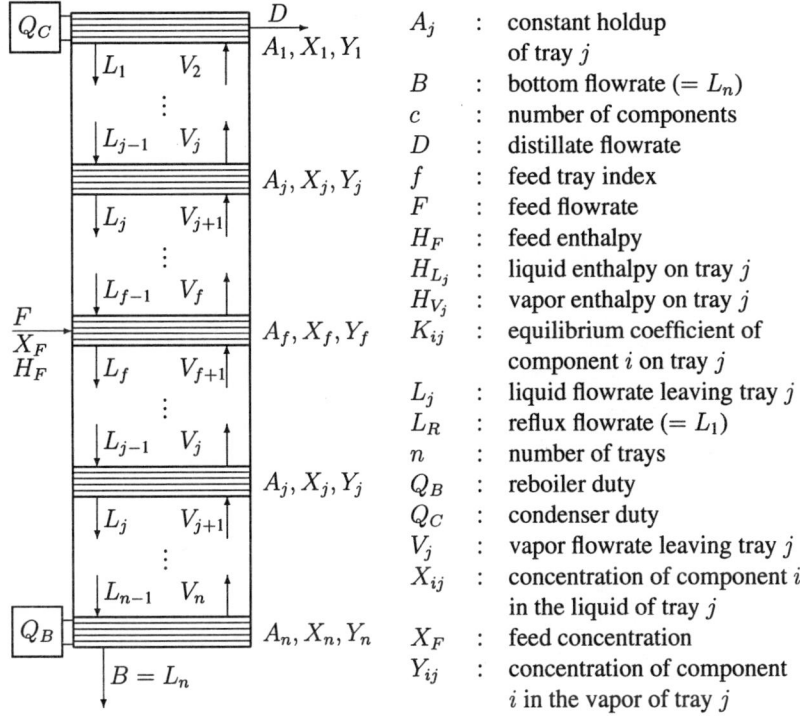

A_j	:	constant holdup of tray j
B	:	bottom flowrate ($= L_n$)
c	:	number of components
D	:	distillate flowrate
f	:	feed tray index
F	:	feed flowrate
H_F	:	feed enthalpy
H_{L_j}	:	liquid enthalpy on tray j
H_{V_j}	:	vapor enthalpy on tray j
K_{ij}	:	equilibrium coefficient of component i on tray j
L_j	:	liquid flowrate leaving tray j
L_R	:	reflux flowrate ($= L_1$)
n	:	number of trays
Q_B	:	reboiler duty
Q_C	:	condenser duty
V_j	:	vapor flowrate leaving tray j
X_{ij}	:	concentration of component i in the liquid of tray j
X_F	:	feed concentration
Y_{ij}	:	concentration of component i in the vapor of tray j

Figure 1.4: Distillation column.

The descriptive equations come from the same source as for the single stage separator presented in the previous section: a thermodynamic model of the treated mixture, material and heat balances.

The balance equations apply for each plate (structure of the process). The material balances concern each component as well as global flows (multicomponent mixture). The equilibrium equations are somewhat more complicated than in the binary case.

The following assumptions have been used when building the model:

- on each plate, the liquid phase and the vapor phase are homogeneous, and in thermodynamic equilibrium,

- the heat losses are neglected,

- the vapor holdup is neglected,

- the liquid holdup is constant,

- the pressure is perfectly regulated.

This is enough to support the following discussion. However, in an actual application, each of these assumptions should be carefully revised.

The following equations are thus obtained:

- Thermodynamic model:

$$0 = Y_{ij} - K_{ij}X_{ij}, \quad i = 1,\ldots,c, \quad j = 1,\ldots,n \tag{1.37}$$

$$0 = \sum_{i=1}^{c} Y_{ij} - \sum_{i=1}^{c} X_{ij}, \quad j = 1,\ldots,n \tag{1.38}$$

$$0 = K_{ij} - k_i(P_j, T_j, X_j, Y_j), \quad i = 1,\ldots,c, \quad j = 1,\ldots,n \tag{1.39}$$

$$0 = H_{L_j} - \sum_{i=1}^{c} X_{ij}hl_i(P_j, T_j) + \Delta H(P_j, T_j, X_j, Y_j), \quad j = 1,\ldots,n \tag{1.40}$$

$$0 = H_{V_j} - \sum_{i=1}^{c} Y_{ij}hv_i(P_j, T_j) + \Delta H(P_j, T_j, X_j, Y_j), \quad j = 1,\ldots,n \tag{1.41}$$

where X_j means $X_{ij}, \quad i = 1,\ldots,c$ and Y_j means $Y_{ij}, \quad i = 1,\ldots,c$.

- Material balances:
  For $i = 1,\ldots,c$:

$$A_1\dot{X}_{i1} = V_2(Y_{i2} - X_{i1}) \tag{1.42}$$

$$\vdots \tag{1.43}$$

$$A_j\dot{X}_{ij} = L_{j-1}X_{i\,j-1} - L_jX_{ij} - V_jY_{ij} + V_{j+1}Y_{i\,j+1}, \tag{1.44}$$

$$j = 2,\ldots,f-1$$

$$\vdots \tag{1.45}$$

$$A_f\dot{X}_{if} = L_{f-1}X_{i\,f-1} - L_fX_{if} - V_fY_{if} + V_{f+1}Y_{i\,f+1} + FX_F \tag{1.46}$$

$$\vdots \tag{1.47}$$

$$A_j\dot{X}_{ij} = L_{j-1}X_{i\,j-1} - L_jX_{ij} - V_jY_{ij} + V_{j+1}Y_{i\,j+1}, \tag{1.48}$$

$$j = f+1,\ldots,n-1$$

$$\vdots \tag{1.49}$$

$$A_n\dot{X}_{in} = L_{n-1}X_{i\,n-1} - L_nX_{in} - V_nY_{in} \tag{1.50}$$

- Thermal balances:

$$A_1\dot{H}_{L_1} = V_2(H_{V_2} - H_{L_1} - Q_C) \tag{1.51}$$

$$\vdots \tag{1.52}$$

$$A_j\dot{H}_{L_j} = L_{j-1}H_{L_{j-1}} - L_jH_{L_j} - V_jH_{V_j}Y_{ij} + V_{j+1}H_{V_{j+1}}, \tag{1.53}$$

$$j = 2,\ldots,f-1$$

$$\vdots \tag{1.54}$$

$$A_f \dot{H}_{L_f} = L_{f-1}H_{L_{f-1}} - L_f H_{L_f} - V_f H_{V_f} + V_{f+1}H_{V_{f+1}} + FH_F \quad (1.55)$$

$$\vdots \quad (1.56)$$

$$A_j \dot{H}_{L_j} = L_{j-1}H_{L_{j-1}} - L_j H_{L_j} - V_j H_{V_j} Y_{ij} + V_{j+1}H_{V_{j+1}}, \quad (1.57)$$

$$j = f+1, \ldots, n-1$$

$$\vdots \quad (1.58)$$

$$A_n \dot{H}_{L_n} = L_{n-1}H_{L_{n-1}} - L_n H_{L_n} - V_n H_{V_n} + Q_B \quad (1.59)$$

As for the separator, the enthalpy derivatives are expressed as functions of the concentrations and their derivatives, and equations (1.51) to (1.59) reduce to algebraic ones.

Hereafter, the emphasis will be put on various aspects of the simplification of this system of equations and on the structure reduction. Meanwhile, the problem of resolution will not be treated.

1.4.2 Reduction of the computing time, simplification of the physical laws

An analysis of the repartition of the computing time shows that the most important part is played by the thermodynamic model. The computing time associated with the vapor-liquid equilibrium resolution often represents more than 95 % of the total computing effort. It directly influences the overall performance.

The equilibrium laws k_i are not explicit, and come in general from the resolution of nonlinear systems of equations. This complexity is due to the generality of the thermodynamic models used. These models usually cope with large families of components, and large domains of pressures, temperatures and concentrations. Such models – which are themselves knowledge models – can be used for a wide range of processes.

The full range of operation of a distillation column lies in a very restricted part of the domain of application of such models. Therefore, for a given process in this restricted domain, one can postulate the obtention of simpler representations of the thermodynamic properties without a significant loss of accuracy.

For the considered process, the significant components and the bounds of the operating domain are assumed to be known: they come from the specification of the problem and from a basic knowledge of the process.

We assume that a "complete" model, which uses the initial thermodynamic submodel, is available. By running such a model it is possible to refine the actual operating domain and to show a submanifold in a neighborhood of which the states of the system will lie.

The equations representing the equilibrium on the plates are independent from each other. The index referring to the plate number is removed in the following notations.

A central difficulty comes from the dependence of the equilibrium coefficients K_{ij} on the concentrations. To eliminate this dependence, the first solution often used consists of the following approximation:

$$k_{ij}(P, T, X, Y) \simeq \bar{k}_i(P_j, T_j) = k_i(P_j, T_j, X_0, Y_0) \quad (1.60)$$

where X_0 and Y_0 are average values taken for the liquid and vapor concentrations.

This approximation already represents, through the choice of X_0 and Y_0, an adaptation to the particular process under study. Figure 1.5 shows the result obtained for a column with 15 trays treating a 10-component mixture. The dashed curve represents the approximation of the equilibrium coefficient of the component 4 - a key-component. The points show the values given by the initial thermodynamic model, for states distributed in the operating domain.

The above approximation is based on basic information on the range of evolution of the system. We will call it an *a priori* approximation.

By using the complete model, one can note that the variables P, T, X do not behave independently: the state remains in the neighborhood of a submanifold described by equations of the following type:

$$X = \varphi(P, T)$$
$$Y = \psi(P, T)$$

(1.61)

Figure 1.6 shows the dependence of X_4 on the temperature, all trays considered together.

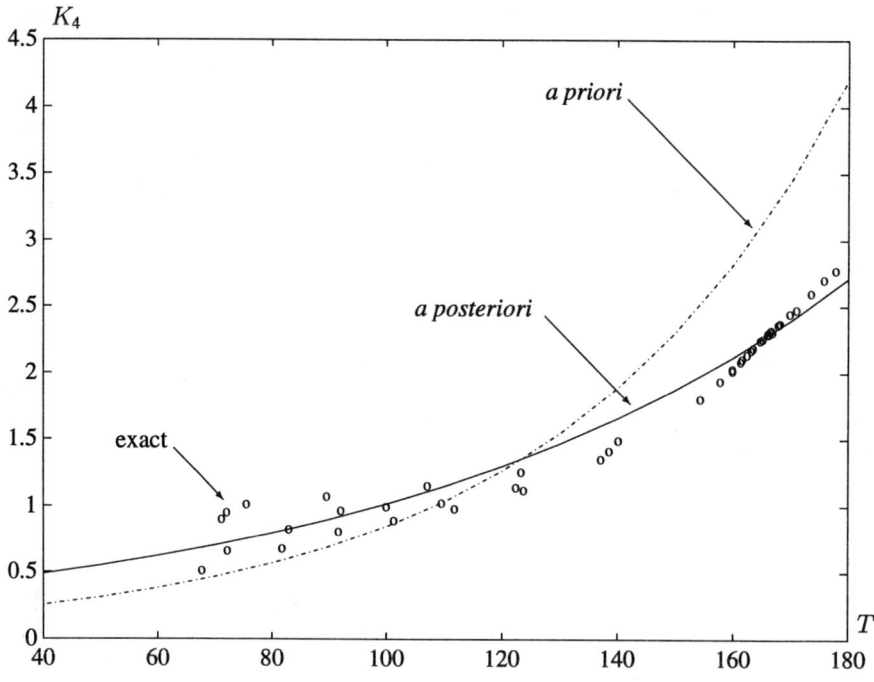

Figure 1.5: Equilibrium coefficients: *a priori* and *a posteriori* approximations.

This suggests a better approximation of the equilibrium coefficients by a law of the following type:

$$k_{ij}(P, T, X, Y) \simeq \tilde{k}_i(P, T) = k_i(P, T, \varphi(P, T), \psi(P, T))$$

(1.62)

This approximation $\tilde{k}$ which uses information coming from the complete model will be called *a posteriori*. The *a priori* approximation, $\bar{k}$, does not use this information.

Figure 1.5 compares the *a priori* and *a posteriori* approximations – dashed and solid lines respectively. The improvement due to the *a posteriori* approximation is clearly evidenced.

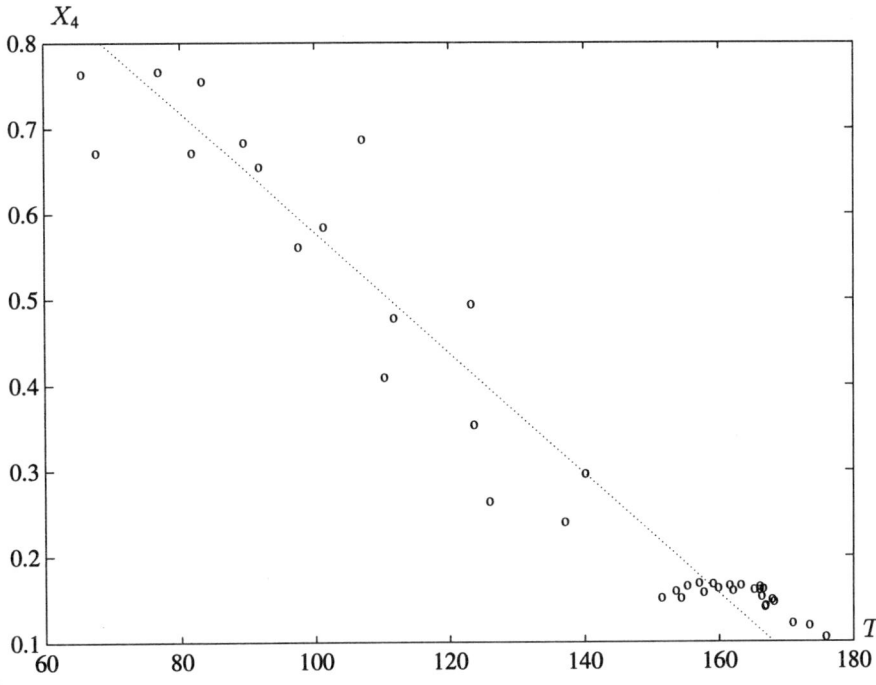

Figure 1.6: Concentration-temperature submanifold.

The procedure described above outlines the principle and the interest of an *a posteriori* approximation. In fact, approximating ψ through regression and substituting them in k does not reduce the complexity of the equations. In practice, it is not necessary to explicitly model φ and ψ. k can be expressed through a direct regression made on a set of states (P, T, X, Y) coming from the complete model and distributed in the whole operating domain. Notice that these data implicitly match the constraints φ and ψ.

In the case of refinery columns (Braoudakis,1975), the error induced by this *a posteriori* simplification procedure was found to be much smaller than the uncertainty of the original thermodynamic model – from Chao Seader techniques – while the computing time was reduced by a ratio of more than 20.

Remark 7

- *The same approach is adopted for the representation of the enthalpies, (equations (1.40) and (1.41)).*

- *This step of simplification modifies only the detail of the representation of the thermodynamic model. The general structure of the model of the column remains unchanged.*

1.4.3 Aggregation

The number of states of the model, which remains unchanged after the previous step of simplification, is often too large. An aggregation step is then necessary. Two directions will be followed: the reduction of the number of components, and the reduction of the number of trays.

Here, we are dealing with "physical" aggregation: The aggregate model will remain the model of a distillation column, but with a simpler structure. The internal coherence and the main properties of the model will remain unchanged.

1.4.3.1 Component aggregation

Reducing the number of components is useful when the treated mixture is complex. This is the case, for instance, for crude oil distillation. Crude oil is usually described as a set of more than 100 main components, the higher ones being themselves aggregates of similar single molecular species.

The approach presented will again put in evidence the concept of *a priori* and *a posteriori* approximation. For a given column and for a given class of crude mixtures, it will be possible to make a drastic aggregation within the constraint of an acceptable loss of accuracy.

Consider a subset of components $\{i = k_1, \ldots, k_2\}$ and assume that we want to represent it as a single equivalent pseudocomponent. The following relations should hold:

$$
\begin{aligned}
\hat{X} &= \sum_{i=k_1}^{k_2} X_i \\
\hat{Y} &= \sum_{i=k_1}^{k_2} Y_i \\
&= \sum_{i=k_1}^{k_2} k_i(P,T)X_i \\
\bar{k}(P,T,W) &= \hat{k}(P,T,X) \\
&= \frac{\sum_{i=k_1}^{k_2} k_i(P,T)X_i}{\sum_{i=k_1}^{k_2} X_i}
\end{aligned}
\tag{1.63}
$$

where $\hat{X}, \hat{Y}, \hat{k}$ are the parameters characterizing the liquid-vapor equilibrium for the pseudocomponent, and where $W_i = X_i / \sum_{j=k_1}^{k_2} X_j$, $i = k_{1,\ldots,k_2}$ (W is the normalized "internal" concentration of the pseudocomponent).

In order to use a pseudocomponent as an ordinary component in the model of the column, we need to remove the dependence of $\bar{k}$ on W. The first solution consists of setting:

$$
\tilde{k}(P,T) = \bar{k}(P,T,W_0)
$$

for a fixed reference W_0 of W. This is the *a priori* procedure, already used for obtaining the original aggregates. By using of the complete model, one can again define a submanifold

$W = \varphi(P,T)$ in the neighborhood of which the internal concentrations of the pseudo-components remain while the whole domain of operation of the column is explored. The approximation is given by:

$$\tilde{k}(P,T) = \bar{k}(P,T,\varphi(P,T))$$

The use of this approach to simplify the model of an atmospheric crude distillation column allowed a reduction from 70 to 10 pseudocomponents.

1.4.3.2 Tray aggregation

A second way to reduce the size of a column model consists in reducing the number of plates. Alvarez (1979) proposed the aggregation of a subset of trays into an equivalent plate whose parameters are obtained from the initial model through a minimization procedure. This approach was successfully applied to a refinery debutanizer.

Another approach, proposed by Rouchon (1990) consists of:

- concentrating the liquid holdups in a reduced number of trays,
- transforming the differential equations into the corresponding steady state algebraic equations for the remaining trays.

Excellent results were obtained on a refinery depropanizer and the simplified model was used to implement a nonlinear control law.

1.4.4 Bilinear modeling

Consider a binary column for which the heat balances are replaced by the Lewis assumption – constant flowrates on each section of the column. If, on each tray, the equilibrium is approximated, for one selected component, by a linear law, then the model is bilinear.

It is possible to mix this approach with the tray aggregation described in the previous section, obtaining a low order bilinear model. This approach was developed and validated on a pilot plant (Espana,1977).

To conclude this progressive approach to model simplification, it is possible to elaborate a complete family of approximated models whose complexity decreases from the complete original model to bilinear low order models with experimentally adapted parameters.

1.5 Adaptation

The theorical laws used in knowledge models are seldom known with enough accuracy in order to develop a model which is sufficiently accurate for an actual application. It is then necessary to "adapt" the model by using a set of experimental data.

The example of distillation will support a discussion of the choice of the number and location of the adaptation parameters, in relation to the available experimental data.

1.5.1 Example: industrial distillation column

The structure of the column under study is equivalent to that given in figure 1.4. It belongs to the low temperature distillation section of a steam cracker, see Bornard (1971).

A knowledge model is given by equations (1.37) to (1.59), and experimental data were collected, from which steady state periods of operation were selected.

Two sources of information are available – the knowledge model and the experimental data – which are partially uncertain and possibly contradictory. The model contains the effects of limited knowledge and of various simplifying assumptions. The experimental data set is corrupted by the measurement noise and biases, and by the disturbances occurring during the experiments.

The adaptation of the model – combining these two information subsets – requires the comparison of the various uncertainties appearing in both of them.

The variables experimentally measured or estimated are the following:

- pressure P,
- flowrates F, D, B, L_R,
- concentrations X_{F_i}, $i \neq 3$, X_{n3}, X_{14},
- enthalpy H_F,
- temperatures T_1, T_n, T_k.

1.5.2 Coherence of the experimental data

Among the equations of the system, the conservation laws have their certainty level higher than that of the involved measurements. For this subset of equations and measured variables, the model should be used for improving the quality of the experimental data set.

Consider the global material balance equations:

$$F = D + B \tag{1.64}$$
$$FX_{f_i} = DX_{i1} + BX_{in} \quad i = 1, \ldots, c \tag{1.65}$$
$$\sum_{i=1}^{c} X_{i1} = 1 \tag{1.66}$$
$$\sum_{i=1}^{c} X_{in} = 1 \tag{1.67}$$

The separation performed in the column is high enough for components other than the key ones (3 and 4) to be fully separated in practice. Thus:

$$\begin{aligned} X_i &\simeq 0 \quad i = 5, \ldots, c \\ X_i &\simeq 0 \quad i = 1, 2 \end{aligned} \tag{1.68}$$

The equations (1.64) to (1.68) are equivalent, after elimination, to two relations between the measured variables F, D, B, X_{F_i}, $i \neq 3$, X_{n3}, X_{14}. These equations will necessarily be satisfied by the model. Therefore, one should elaborate, from the rough measurement set, a corrected set satisfying these equations. This operation is performed by projection

– weighted least squares. Caujolle (1976) gives a detailed analysis of the elaboration of coherent data sets from experimental data and constraint equations coming from a model.

Once this task is achieved, a selection is made among the variables in the corrected set, in order to obtain a minimal independent data set.

1.5.3 Choice of the adaptation parameters

The number of variables exceeds by 4 the number of equations. One can use this information to improve the model. This can be done by:

- "adapting" the value of certain physical parameters, to be chosen in the poorly known ones,
- introducing "adaptation parameters" in certain equations in order to take into account neglected phenomena.

The parameters to be adapted should satisfy number and sensitivity conditions:

- their number cannot be greater than the number of experimental measured variables in excess,
- the parameters should be distinguished by the measured variables.

One touches here the problem of identifiability (see Chapter 3), under both its structural and its sensitivity aspects.

For the column under study, the prediction of the equilibrium coefficient and the inperfect operation of the trays constitute the major sources of uncertainty of the model. The adaptation should take place in this area.

A favorable situation is when a parameter can be estimated from a data subset, independently from the other parameters and data. This is practically the case for the top and bottom equilibria. For each of these two trays, we know the pressure, temperature and liquid concentrations (solution of the equations (1.64) to (1.67) under the approximation (1.68)). These variables are linked by the equilibrium equations. For each tray there is one excess equation with respect to the unknown variables, and one adaptation parameter can be associated to each. A multiplicative factor on the equilibrium coefficients changes the equilibrium temperature – pressure relation without affecting the separation between light and heavy components. Two parameters e_3 and e_4 are defined in this way.

These parameters constitute an adaptation of the thermodynamic model. The magnitude of this adaptation is known at the top and at the bottom of the column. Without extra information, an interpolation is used to set the value of the same parameter on each tray. Moreover here the question of the interpolation variable arises: temperature, pressure, and tray position could be candidates. The latter has the advantage of simplicity, although it may not be the best option.

The overall separation factor which characterizes the efficiency of the operation must imperatively be adapted. This property is strongly linked to the efficiency of the trays. The model was extended through the introduction of a classical Murphree efficiency parameter.

A last redundant measurement remains unused: the temperature of a "sensitive" tray. It is convenient that the model gives a good representation of this temperature since it is used in many control schemes.

Finally two efficiency parameters e_1, and e_2, are retained, for the top and the bottom sections of the column respectively, since the flows in the two sections are quite different.

The maximum number of parameters is then reached. This is not a general recommendation but only the choice made in this particular case, after the analysis which is briefly described above. In particular, it is convenient to check that the number of parameters taken is not excessive, leading to unacceptable margin of uncertainty and dispersion on these parameters.

1.5.4   Results, validation

A value of the parameter was thus obtained for each steady state. Figure 1.7 shows the results obtained for e_1 and e_2. Their low dispersion is a good indication of the parameter significance.

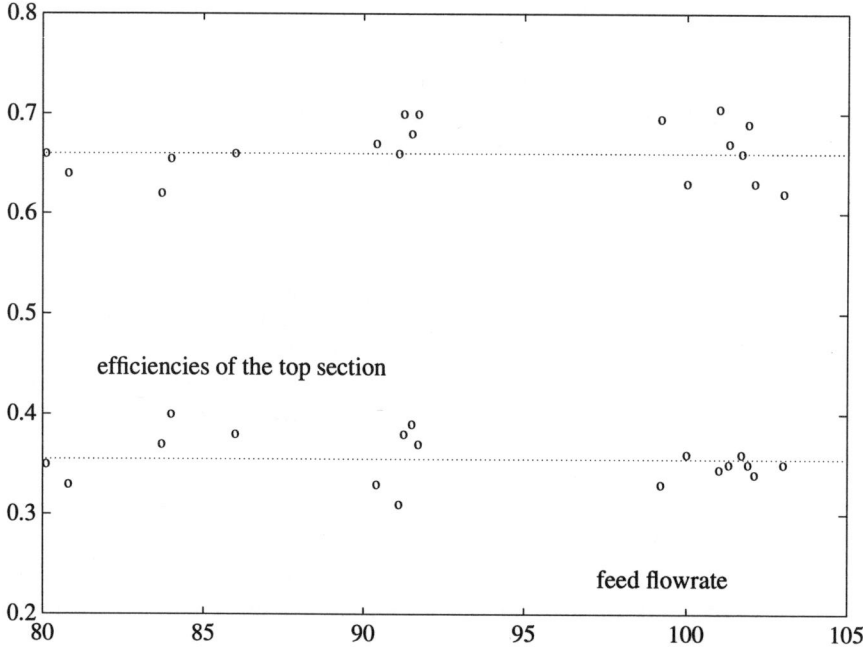

Figure 1.7: Identified efficiencies as a function of the feed flowrate.

The computed parameters should be independent of the operating point. If this is not the case, a correlation is to be found with some of the descriptive variables in order to take advantage of the remaining unused information.

In the example, it was not possible to find a significant dependence of the identified parameters either on the feed flowrate – see figure 1.7 – or on other variables.

The parameters e_3 and e_4 behave similarly.

It is important to validate the model on an experimental data set independent of the set used for the adaptation. In the example, independent test runs were performed after the adaptation stage. The maximum error between the measured and predicted variables was 2 % on the reflux flowrate and 1 °C on the temperatures.

1.6 Conclusions

This chapter presented various aspects of the modeling task. The main stages of modeling were pointed out, as well as typical difficulties that arise during applications, and the principles which allow us to overcome them. A presentation in the form of small case studies gave an idea of the diversity of the problems which may be encountered.

The engineer in charge of a modeling project should at the same time:

- collaborate with the process specialist for the choice of the theoretical models,
- specify the required performances and constraints as a function of the foreseen use of the model,
- obtain enough information both on the internal behavior of the model and on the measurement system accuracy, in order to make an adequate parametrization,
- carefully follow the stages of numerical resolution.

All these aspects are complementary to the unifying design approach which is presented in Chapter 2.

1.7 Bibliography

[1] ALVAREZ J. (1979). *Modélisation simplifiée des colonnes de distillation par agrégation de plateaux et applications*, Thesis, INPG, Grenoble.

[2] BENALLOU A., D.E. SEBORG AND D.A. MELLICHAMP (1986). Dynamic compartmental models for separation processes, *A.I.Ch.E.J.*, 32, 1067–1078.

[3] BORNARD G. (1971). *Contribution à l'étude des colonnes à distiller*, PhD Thesis, USMG, Grenoble.

[4] BORNARD G. (1976). *Modeling of industrial processes*, Lecture Notes, LAG, Grenoble.

[5] BORNARD G., G. BRAOUDAKIS AND J.P. GAUTHIER (1979). *Dynamic simulation of multicomponent distillation columns with variable pressure. Application to a refinery debutanizer*, Congres IMACS, Sorrento.

[6] BORNARD G. AND J.P. GAUTHIER (1981). *Modélisation dynamique des colonnes de distillation*, Outils et modèles mathématiques pour l'Automatique et le traitement du signal, I.D. Landau ed., vol. 1, Editions du CNRS.

[7] BORNARD G. AND R. PERRET (1977). *About some modeling and control problems in petrochemical industries*, Symposium IFAC-DCAPC, Delft.

[8] BRAOUDAKIS G. (1975). *Modélisation des colonnes à distiller et application à la simulation*, Thesis, INPG, Grenoble.

[9] BRENT R.P. (1973). Some efficient algorithms for solving systems of nonlinear equations, *SIAM J. Numer. Anal.*, 10, 327–344.

[10] CAUJOLLE J.P. (1977). *Amélioration d'un ensemble de mesures brutes industrielles par traitement de cohérence matérielle et thermique*, Thesis, INPG, Grenoble.

[11] ESPANA M. (1977). *Modélisation bilinéaire des colonnes à distiller*, Thesis, INPG, Grenoble.

[12] FOULARD C. AND G. BORNARD. (1973). *Identification and parameter estimation of distillation columns, a case study*, 3rd symposium IFAC on identification and parameter estimation, The Hague.

[13] FRIEDLY J.C. (1972). *Dynamic behavior of processes*, Prentice Hall, Englewood Cliffs.

[14] GEAR C.W. (1971). *Numerical initial value problems in ordinary differential equations*, Prentice Hall, Englewood Cliffs.

[15] GILLES G. (1971). *Élaboration du modèle mathématique et commande optimale par calculateur numérique d'un échangeur thermique pilote*, Doctorate Thesis, USMG, Grenoble.

[16] HALL D.M. AND WATTS J.M. (1976). *Modern methods for ordinary differential equations*, Oxford University Press, Oxford.

[17] HIMMELBLAU D.M. AND BISCHOFF K.B. (1968). *Process analysis and simulation*, Wiley, New York.

[18] JAOUAHRI S. (1980). *Modélisation statique et dynamique d'un échangeur thermique à quatre fluides*, Internal note LAG, Grenoble.

[19] MORÉ J.J. AND M.Y.COSNARD (1979). Numerical solution of nonlinear equations, *ACM Trans. Math. Softw.*, 5, 1, 64–68.

[20] PETZOLD L.R. AND P. LÖTSTED. (1986). Numerical solution of nonlinear differential equations with algebraic constraints (II) – Practical implications, *SIAM J. of Scientific and Statistical Computation*, 7, 720–733.

[21] RICHALET J, RAULT AND R. POULIQUEN. (1971). *Identification des processus par la méthode du modèle*, Gordon and Beach, Paris.

[22] ROUCHON P. (1990). *Simulation dynamique et commande non linéaire des colonnes à distiller*, PhD Thesis ENSMP, Paris.

[23] WOLFE M.A. (1978). *Numerical methods for unconstrained optimization*, Van Nostrand Reinhold.

CHAPTER 2

Bond-graph modeling of physical systems

G. DAUPHIN-TANGUY, S. SCAVARDA

The bond-graph tool defined by Paynter (1961), formalized by Karnopp (1983), Rosenberg (1990), Thoma (1991), Breedveld (1984), is situated halfway between the physical system and the associated mathematical models (transfer matrix in the linear case, linear or nonlinear state equation, second order differential system equations).

It must be pointed out immediatly that this technique has no pretention to be universal. However, it has shown its efficiency in a large number of examples, just as well for system design, as for simulation and determination of control laws, and therefore, can be ranked among the classical methods for model construction.

As for the design of a model of knowledge, the bond-graph methodology needs the analysis of the physical phenomena which will be taken into account in the modeling procedure (gravity, friction, inertia, compressibility, etc.).

On the other hand, we can show a certain number of fundamental differences.

Firstly, the bond-graph approach does not need to explicitly write the general conservation laws. It lies essentially on the characterization of power exchange phenomena within the system. The procedure is composed of several steps.

The first step consists of the study of the system architecture, either the interaction of the components or the coupling of the retained physical phenomena, and its combination with a unique graphical language for the whole set of physical domains. The bond-graph model can also be obtained by interpreting the discretization methods in mechanics of structures (finite elements, modal analysis, modal synthesis, etc.).

The second step consists of the writing of the characteristic laws of components or phenomena, linear or nonlinear. In this way, the concept of causality is a major asset of this technique. Indeed, the bond-graph model allows us to show the cause and effect relations in the system and guides, in a systematic manner, the writing and the organization of equations.

The mathematical models obtained are then linear or nonlinear, and the property of non-linearity of the model can be identified as due to the structure and/or to the components, which may appear to be of great interest for introducing simplifications from partial linearization.

The bond-graph can evolve, for a more accurate modeling, simply by adding new bond-graph elements, without any need to undertake the whole procedure from the beginning.

The particular choice of state variables, which are always associated with a component from the system or a remarkable physical phenomenon, gives to the thus obtained state variables a non neglectible physical insight.

Also, due to its graphical character and causal structure, the bond-graph model appears to be an excellent analysis tool. The definitions of causal path and causal loop lead to some very interesting results.

It is thus possible, while running through the bond-graph and following the causality, to derive the associated block-diagram, even with nonlinear blocks, which could be very interesting for simulation.

Likewise, and just by considering the causal connexions between elements, we can derive:

- directly the state equations of the system;
- information concerning the dynamic variation domains of the system as well as the corresponding variables;
- the simplified bond-graph;
- information about the structural properties of the system.

The first part of this chapter presents the basic tools for bond-graph modeling.

In the second part, the notion of causality and the causal properties of a bond-graph model are developed. This graphical representation of cause to effect relations constitutes the major interest of the bond-graph approach and its superiority over other graph representations.

In the third part, two applications are discussed, one for the modeling of power electronics converters, the other for the modeling of the power part of an electropneumatic actuator.

2.1   Basic tools of the bond-graph modeling

This part presents succinctly the main principles and the basic tools of the bond-graph modeling.

After defining the graphical symbols associated with power transfer, with physical phenomena coding, and with the definition of the generalized variables in the system, the classical procedures to construct a bond-graph model in the mechanical, electrical and hydraulical domains are detailed.

2.1.1   Power transfer representation

Consider the two subsystems, one mechanical (a) and the other electrical (b) represented in figure 2.1.

In both cases, a physical linkage between A and B exists, either thanks to a coupling bar (supposed to be rigid and massless) in case (a), or using two electrical wires (supposed lossless) in case (b). In the closed system, composed of the two subsystems A

and B, there is not only energy conservation but also power continuity (input power − output power = 0).

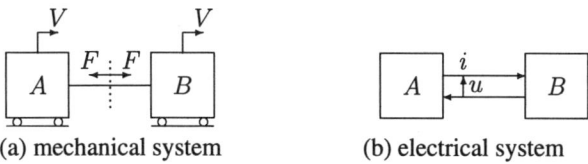

(a) mechanical system   (b) electrical system

Figure 2.1: Physical diagrams.

The power flow between A and B is represented by a power bond, characterized by the following symbol:

The instantaneous power exchanged between A and B is determined as $P = FV$ in mechanics or $P = ui$ in the electricity domain. The bond carries the variables which intervene in the power calculus and the half-arrow direction corresponds to the positive direction of the power and the flow variables.

The physical schemes of figure 2.1 have then their bond-graph translation in figure 2.2.

$$A \xrightarrow[\quad V \quad]{\quad F \quad} B \qquad\qquad A \xrightarrow[\quad i \quad]{\quad u \quad} B$$

(a) mechanical system   (b) electrical system

Figure 2.2: Power transfer associated with diagrams 2.1.

2.1.2 Introduction of the different variables

2.1.2.1 Power variables

We have seen before that the exchanged power P can be expressed as the product of two complementary variables u (or F) and i (or V). From a general point of view and independently of the considered domain, these variables are called "generalized" variables of **effort** and **flow** and noted respectively e and f. They are **power variables**, and we have:

$$P = ef. \tag{2.1}$$

As a convention, the bond is always represented with its half-arrow drawn pointing to the flow variable:

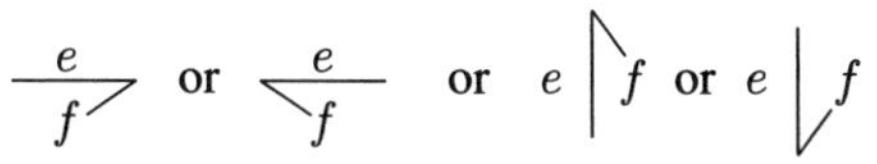

$$\underset{f}{\overset{e}{\longrightarrow}} \quad \text{or} \quad \underset{f}{\overset{e}{\longleftarrow}} \quad \text{or} \quad e \Big\uparrow f \quad \text{or} \quad e \Big\downarrow f$$

This leads, in the mechanical and electrical domains, to:

$$\xrightarrow[\;i\;]{\;u\;} \qquad \xrightarrow[\;V\;]{\;F\;}$$

2.1.2.2 Energy variables

Energy is defined as the integral of power with respect to time:

$$E(t) = \int_0^t P(\tau)d\tau, \qquad E(0) \quad \text{supposed null.} \tag{2.2}$$

The **energy variables** are defined using the following integral relationships:

$$
\begin{aligned}
p(t) &= \int_0^t e(\tau)d\tau, \qquad p(0) \quad \text{supposed null,}\\
q(t) &= \int_0^t f(\tau)d\tau, \qquad q(0) \quad \text{supposed null,}
\end{aligned}
\tag{2.3}
$$

$p(t)$ is the **generalized moment** and $q(t)$ the **generalized displacement**.

The following table gives an indication about the meaning of these generalized variables for some physical domains.

domains	effort e	flow f	momentum p	displacement q
mechanical:				
translation	force F	velocity V	momentum p	displacement x
rotation	torque τ	angular velocity ω	angular momentum h	angle θ
electrical	voltage u	current i	flux linkage Φ	charge q
hydraulical	pressure P	flow rate Q	pressure impulse Pp	volume V

Table 2.1: Generalized energy variables significance for different physical domains.

2.1.3 Bond-graph elements

The bond-graph elements can be classified as passive elements, active elements and junction elements.

2.1.3.1 Simple passive elements

These elements are said to be **passive** since they **transform the received power** into stored or dissipated energy. The **half-arrow** is always represented as **entering into these elements**.

2.1.3.1.1 *R-element*

The *R*-element is used to model any phenomenon which statically links effort and flow variables.

As examples, we can speak of a mechanical damper or dashpot, an electrical resistor, a diode, an hydraulic restriction, etc.

It characteristic generic law is:

$$\Phi_R(e, f) = 0. \tag{2.4}$$

This law may be linear or nonlinear, and in the linear case, for instance, it can be written:

$$u = R_1 i \quad \text{or} \quad F = bV.$$

The general representation is:

$$\xrightarrow[\ f\ ]{\ e\ } R$$

and in the electrical and mechanical domains becomes:

$$\xrightarrow[\ i\ ]{\ u\ } R : R_1 \qquad \xrightarrow[\ V\ ]{\ F\ } R : b$$

R represents thus the type of the identified phenomenon and R_1 or b the parameter's value which appears while writing the linear law. In the nonlinear case, it is generally hard to specify the law directly on the scheme. Sometimes it is possible to indicate how the parameter varies. For a hydraulic restriction of a servo-valve, for example, when $R :$ $R(x_d)$ is specified, this indicates that Bernouilli's law, which expresses the relationship between the flow rate in the restriction and the difference of pressure between the two extremities of the restriction, depends on the position, x_d, of the cylinder.

R is an energy dissipative element.

2.1.3.1.2 *C-element*

The *C*-element is used to model any physical phenomenon which statically links the effort and displacement variables.

A spring, an accumulator, an electrical capacitor, a storage tank and any elasticity or compressibility phenomenon correspond to that definition.

Its characteristic generic law is:

$$\Phi_C(e, q) = 0, \tag{2.5}$$

which leads, for example, in the linear case, to:

$$u = \frac{1}{C_1} \int i \, dt \quad \text{or} \quad u = \frac{q}{C_1} \quad \text{in the electrical field,}$$

$$F = k \int V \, dt \quad \text{or} \quad F = kx \quad \text{in the mechanical field.} \tag{2.6}$$

The general representation of a C-element is:

$$\xrightarrow{\quad e \quad} C$$
$$f = dq/dt$$

and in the electrical and mechanical domains becomes:

$$\xrightarrow{\quad u \quad} C : C_1 \qquad \xrightarrow{\quad F \quad} C : 1/k$$
$$i = dq/dt \qquad\qquad V = dx/dt$$

C is an energy storage element. This energy is calculated using the power as follows:

$$E(t) = \int_0^t e(\tau).f(\tau) \, d\tau + E_0 \tag{2.7}$$

E_0 is the energy stored initially at $t = 0$ (if any).
Writting $e = e(q)$ and $f \, dt = dq$ leads to the following equivalent form, expressed in terms of the energy variable q:

$$\mathcal{E}(q) = \int_{q_0}^q e(\eta) d\eta + E_0. \tag{2.8}$$

E_0 is the energy stored when $q = q_0$. Usually, it is convenient to define the energy stored to be zero when the effort is zero. So, q_0 will correspond in the following to that value of q at which $e = 0$, and so $E_0 = 0$.

Let us notice, in the linear case, that the electrical energy stored in a capacitor can be written as:

$$\mathcal{E}(q) = \int_{q_0}^q u(\eta) d\eta = \int_{q_0}^q \frac{\eta}{C_1} \, d\eta = \frac{(q^2 - q_0^2)}{2C_1}, \tag{2.9}$$

where q represents thus the electrical charge.

Likewise, the potential energy stored in a spring is expressed by:

$$\mathcal{E}(x) = \int_{x_0}^x F(\eta) \, d\eta = \int_{x_0}^x k\eta \, d\eta = \frac{k(x^2 - x_0^2)}{2}. \tag{2.10}$$

2.1.3.1.3   I-element

The I-element is used to model any phenomenon which statically links the flow and moment variables.

A translating mass, a rotating inertia and an inductance can be mentionned.

Its characteristic generic law is:

$$\Phi_I(p, f) = 0, \tag{2.11}$$

which leads, for example in the linear case, to:

$$u = L\frac{di}{dt}, \quad \text{or} \quad \phi = Li,$$

in the electrical field, and

$$F = M\frac{dV}{dt}, \quad \text{or} \quad p = MV,$$

in the mechanical field.

The general representation of an I-element is:

$$\frac{e = dp/dt}{f} \searrow I$$

and in the electrical and mechanical domains becomes:

$$\frac{u = d\phi/dt}{i} \searrow I : L \qquad \frac{F = dp/dt}{V} \searrow I : M$$

I is an energy storage element. This energy is calculated using the power as follows:

$$E(t) = \int_0^t e(\tau).f(\tau)\,d\tau + E_0 \tag{2.12}$$

with:

$$f = f(p), \quad \text{and} \quad e\,dt = dp, \tag{2.13}$$

which leads to the following equivalent form, expressed in terms of the energy variable p:

$$\mathcal{E}(p) = \int_{p_0}^p f(\eta)\,d\eta + E_0. \tag{2.14}$$

If the energy is defined to vanish when f vanishes and if p_0 coresponds to the point at which $f = 0$, then $E_0 = 0$.

In the linear case, the kinetic energy stored in a mass, for example, can be written:

$$\mathcal{E}(p) = \int_{p_0}^p V(\eta)\,d\eta = \int_{p_0}^p \frac{p}{M}\,d\eta = \frac{(p^2 - p_0^2)}{2M}, \tag{2.15}$$

p represents the momentum of mass M.

Likewise, the magnetic energy stored in a coil can be written as:

$$\mathcal{E}(\Phi) = \int_{\phi_0}^\phi i(\eta)\,d\eta = \int_{\phi_0}^\phi \frac{\eta}{L}\,d\eta = \frac{(\phi^2 - \phi_0^2)}{2L}. \tag{2.16}$$

Remark 1

A confusion may often occur between energy, noted $\mathcal{E}$, and co-energy, which will be noted $\mathcal{E}^*$.

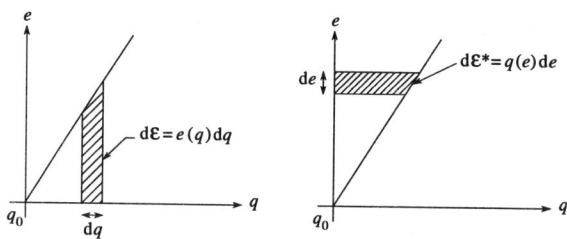

Figure 2.3: Elementary energy and co-energy in a C-element.

In this way, a C-element, relating the variables of effort e and displacement q, permits to define the elementary energy and co-energy, $d\mathcal{E}$ and $d\mathcal{E}^*$, as shown in figure 2.3.

These two quantities are linked by the following relationship (Legendre's transformation), which is represented in figure 2.4 where q_0 is supposed equal to zero:

$$e_1 q_1 = \mathcal{E} + \mathcal{E}^*, \tag{2.17}$$

with:

$$\mathcal{E} = \int_0^{q_1} e(q)dq \quad \text{and} \quad \mathcal{E}^* = \int_0^{e_1} q(e)de.$$

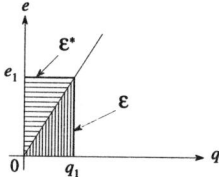

Figure 2.4: Relationship between energy and co-energy in a C-element.

The energy $\mathcal{E}$ is expressed in terms of an energy variable, whereas the co-energy $\mathcal{E}^*$ is expressed in terms of a power variable.

In the linear case, for example a capacitor, the relation $u = q/C$ allows us to calculate $\mathcal{E} = (1/2)(q_1^2/C)$, while the relation $q = Cu$ gives $\mathcal{E}^* = (1/2)Cu_1^2$ and relation (2.17) can be written:

$$u_1 q_1 = \frac{1}{2}\underbrace{\frac{q_1}{C}}_{u_1}.q_1 + \frac{1}{2}\underbrace{Cu_1}_{q_1}.u_1.$$

Likewise, the kinetic co-energy stored in a mass can be expressed, in the linear case by:

$$\mathcal{E}^* = \int_0^{V_1} p(V)dV = \int_0^{V_1} MVdV,$$

and then:

$$\mathcal{E}^* = \frac{1}{2}MV_1^2.$$

2.1.3.2 Active elements: the sources

These elements are said to be active since they supply power to the system. Two kinds of sources can be distinguished:

- effort sources Se (voltage source, gravity force, etc.);
- flow sources Sf (current source, applied speed, etc.).

The **half-arrow** orientation is fixed and supposed **always to be going out of the source element.**

$$Se \longrightarrow \qquad Sf \longrightarrow$$

In each case, one of the two variables (effort or flow) is supposed to be known, and independent of the complementary variable which depends on the system.

2.1.3.3 Junction elements

These elements, noted $0, 1, TF, GY$, serve to interconnect the elements R, C, I, and the sources, and compose the structure junction of the bond-graph model corresponding to the architecture of the studied system. They have the property to be **power conservative.**

2.1.3.3.1 0-junction

This junction allows to associate elements submitted to the same effort.

The property of common effort corresponds, when dealing with simple systems, in the case of unidimensional mechanics, to elements in series (the same force), while in electrical or hydraulical domain to elements in parallel (the same voltage or the same pressure). Consider the following example (fig. 2.5).

We have:

$$e_1 = e_2 = e_3 = e_4 \text{ (characteristic property of the 0-junction),}$$
$$e_1 f_1 + e_2 f_2 - e_3 f_3 - e_4 f_4 = 0 \text{ (power balance),}$$

which leads to:

$$f_1 + f_2 - f_3 - f_4 = 0.$$

The characteristic relations of 0-junction can be expressed as follows:

- equality of efforts for all bonds connected to the junction;
- flow algebraic sum $= 0$.

The weighting of flows occurs using a $(+)$ or $(-)$ sign depending on the direction of the half-arrow, going inside or outside the 0-junction.

2.1.3.3.2 1-junction

This junction allows to interconnect elements submitted to the same flow. It is the dual of the 0-junction.

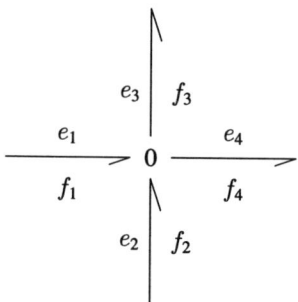

Figure 2.5: 0-Junction.

For unidimensional mechanics, this definition implies the association of simple elements in parallel (the same velocity) while in electrical and hydraulical domains, it implies the association in series (the same current or the same flow rate). By applying these rules to the following example (fig. 2.6) we obtain:

$$f_1 = f_2 = f_3 = f_4 \text{ (characteristic law of the 1-junction)},$$
$$e_1 f_1 + e_2 f_2 - e_3 f_3 - e_4 f_4 = 0 \text{ (power balance)},$$

which leads to:

$$e_1 + e_2 - e_3 - e_4 = 0.$$

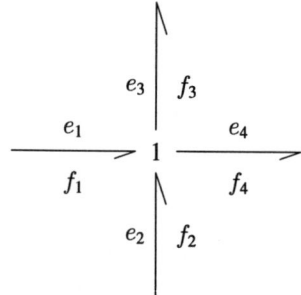

Figure 2.6: 1-junction.

The characteristic relations of the 1-junction can be expressed as follows:

- equality of flow in all bonds connected to the junction;
- effort weighted sum $= 0$.

The weighting of efforts occurs using a $(+)$ or $(-)$ sign depending on the direction of the half-arrow, going inside or outside the 1-junction.

2.1.3.3.3   Transformer TF

This element plays the role of a transducer with power conservation. It is represented by Figure 2.7:

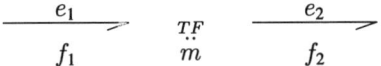

Figure 2.7: Junction TF.

where m is the transformer modulus.

It can be represented as a quadripole (fig. 2.8).

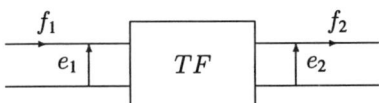

Figure 2.8: Representation of a TF as a quadripole.

Its characteristic relations are:

$$e_1 = me_2,$$
$$f_2 = mf_1.$$
(2.18)

An electrical transformer, a system with pulleys or gears, a lever arm can be mentionned as examples of simple systems modeled by transformers.

This modeling is based on simplification hypothesis which suppose negligible some phenomena such as inertia, friction, heat transfer, etc. which induce power losses are negligible.

When m is not constant, the transformer becomes a modulated transformer and is noted MTF. This element is used, for example, in mechanics (geometric relation between variables, variable reducer, kinematics of mechanisms, etc.).

2.1.3.3.4   Gyrator GY

This transducer element, corresponding also to a quadripole, is represented by:

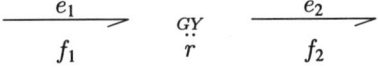

Figure 2.9: Junction GY.

and is characterized by the following relations:

$$e_1 = rf_2,$$
$$e_2 = rf_1,$$
(2.19)

where r is the gyrator modulus. The GY-element crosses the variables, that is the reason for the terminology used to specify it.

The physical components which may be modeled by a GY-element are scarce compared with those for a TF-element. Examples include a simplified gyroscope and a Hall effect sensor.

When r is variable, the gyrator becomes a modulated gyrator noted MGY. This element appears in a mechanics model of the kinematic relations of a solid moving in space.

Remark 2

TF- and GY-elements appear to be of great importance in representing power transformations between various domains. In fact, a TF-element is used to represent the transformation of hydraulic power into mechanical power in an oil cylinder and its modulus is defined as the inverse of the piston's area. Likewise, the electrical power transformation into a mechanical power for a direct-current motor is modeled by a GY-element, the modulus of which is equal to the torque coefficient of the motor.

2.1.4   Procedures for bond-graph models construction

2.1.4.1   Unidimensional mechanical systems

2.1.4.1.1   Procedure

1. Fix a reference frame for velocity orientation.

2. For each distinct velocity establish a 1-junction.

3. Determine the relations linking velocities and represent them using 0-junctions placed between the corresponding 1-junctions.

4. Connect the different junctions to each other using bonds with their half-arrow orientation matching the fixed frame.

5. Place the elements and the sources on their respective junctions.

6. Simplify the bond-graph when possible: nodes with a null velocity are suppressed as well as all bonds connected to them. The following situations are equivalent.

$$\rightharpoonup 1 \rightharpoonup \;\equiv\; \rightharpoonup \qquad \rightharpoonup 0 \rightharpoonup \;\equiv\; \rightharpoonup$$

$$(\text{but not} \;\rightharpoonup 1 \leftharpoonup\; \text{or} \;\rightharpoonup 0 \leftharpoonup)$$

2.1.4.1.2 *Example*

Consider the following system:

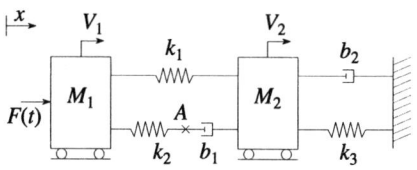

Figure 2.10: Mechanical diagram.

Step 1: Choice of the frame (x direction).

Step 2: Search for the different velocities $\Longrightarrow$ 1-junctions.

Absolute velocities: V_1, V_2 for the masses, V_A.
Velocity (null) of the framework: V_{FW}.
Relative velocities: $V_{k1}, V_{k2}, V_{k3}, V_{b1}, V_{b2}$.

$$
\overset{1}{V_1} \qquad \overset{1}{V_{k_1}} \qquad \overset{1}{V_2} \qquad \overset{1}{V_{b_2}} \qquad \overset{1}{V_{FW}}
$$

$$
\overset{1}{V_{k_2}} \qquad \overset{1}{V_{b_1}} \qquad \overset{1}{V_{k_3}}
$$

$$
\overset{1}{V_A}
$$

Steps 3 and 4: Relations linking absolute and relative velocities $\Longrightarrow$ 0-junctions.

$$
\begin{aligned}
V_{k1} &= V_1 - V_2, & V_{k_2} = V_1 - V_A, & \qquad V_{b_1} = V_A - V_2, \\
V_{b_2} &= V_{k_3} = V_2 - V_{FW}.
\end{aligned}
$$

(2.20)

$$
\overset{1}{V_{k_1}} \qquad\qquad\qquad \overset{1}{V_{b_2}}
$$

$$
\overset{1}{V_1} \qquad\qquad \overset{1}{V_2} \qquad\qquad \overset{1}{V_{FW}}
$$

$$
\overset{1}{V_{k_2}} \qquad \overset{1}{V_A} \qquad \overset{1}{V_{b_1}} \qquad \overset{1}{V_{k_3}}
$$

Step 5: Affectation of elements and sources

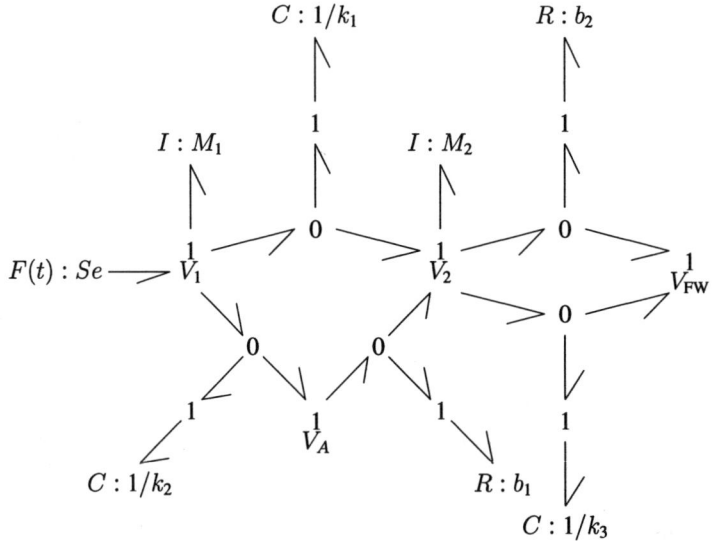

Step 6: Simplifications.

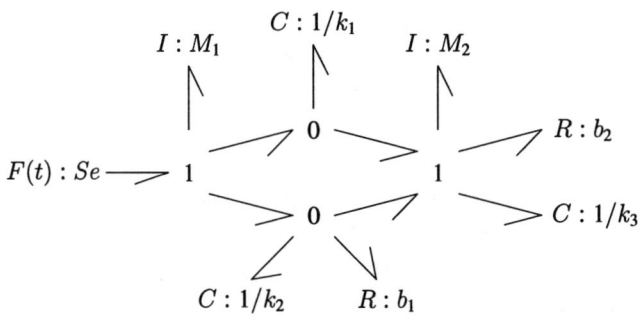

Figure 2.11: Bond-graph model of the mechanical diagram of figure 2.10.

2.1.4.2   Two dimensional dynamics

2.1.4.2.1   *Procedure*

1. Fix a reference frame,

2. Determine the key geometric variables, composed of displacements associated with elasticities (q_c) and inertia (q_I), and (q_k) the displacements corresponding to the generalized coordinate vector in the language of Lagrange mechanics.

3. Write the displacement relations $\begin{cases} q_C = \phi_{Ck}(q_k) \\ q_I = \phi_{Ik}(q_k) \end{cases}$ and differentiate them in order to obtain the velocity relations in the following form:

$$\begin{cases} \dot{q}_C = T_{Ck}(q_k) \cdot \dot{q}_k, \\ v_I = \dot{q}_I = T_{Ik}(q_k) \dot{q}_k. \end{cases}$$

4. Determine junction structure to represent these relations.

5. Connect elements to junctions.

6. Simplify when possible.

2.1.4.2.2 Example

Consider the elastic pendulum represented in figure 2.12 (a). [11]

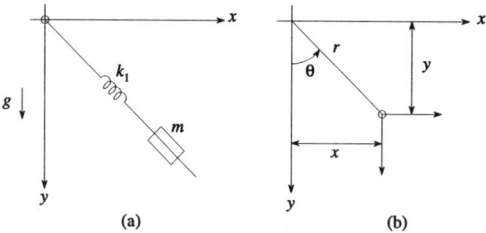

Figure 2.12: The elastic pendulum in plane motion and its key geometric variables.

The key geometric variables are r, θ, x, y, as shown in figure 2.12 (b). The different coordinate vectors can be written:

$$q_c = r, \quad q_I = \begin{pmatrix} x \\ y \end{pmatrix} \quad q_k = \begin{pmatrix} r \\ \theta \end{pmatrix},$$

the geometrical relations between these vectors are:

$$\begin{aligned} q_C &= \phi_{Ck}(r, \theta) = \begin{bmatrix} 1 & 0 \end{bmatrix} \begin{pmatrix} r \\ \theta \end{pmatrix}, \\ q_I &= \begin{pmatrix} x \\ y \end{pmatrix} = \phi_{Ik}(r, \theta) = \begin{bmatrix} r \sin \theta \\ r \cos \theta \end{bmatrix}, \end{aligned} \qquad (2.21)$$

which leads to the following relations between the different velocities:

$$\begin{aligned} \dot{q}_C &= T_{Ck}(r, \theta) \begin{pmatrix} \dot{r} \\ \dot{\theta} \end{pmatrix} \quad \text{with} \quad T_{Ck}(r, \theta) = \begin{bmatrix} 1 & 0 \end{bmatrix}, \\ v_I &= \dot{q}_I = \begin{pmatrix} \dot{x} \\ \dot{y} \end{pmatrix} = T_{Ik}(r, \theta) \cdot \begin{pmatrix} \dot{r} \\ \dot{\theta} \end{pmatrix}, \end{aligned} \qquad (2.22)$$

with:

$$T_{Ik}(r, \theta) = \begin{bmatrix} \sin \theta & r \cos \theta \\ \cos \theta & -r \sin \theta \end{bmatrix}.$$

The $T_{Ik}(r, \theta)$ matrix is modeled in bond-graph by the use of four MTF-elements with non-constant moduli.

Step 4 of the procedure leads to the following bond-graph (fig. 2.13) which represents the kinematic structure of the model.

Step 5, associated with step 6, derives the bond-graph model of the pendulum, represented in figure 2.14.

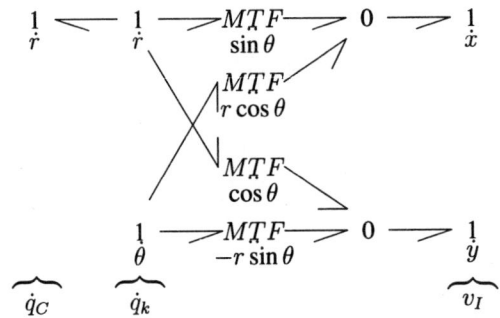

Figure 2.13: Kinematic structure of the pendulum of figure 2.11.

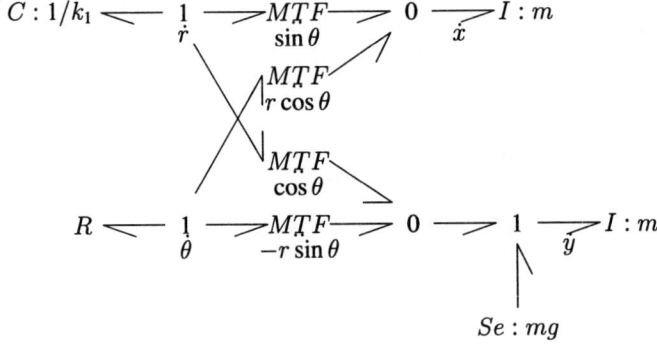

Figure 2.14: Bond-graph associated with the elastic pendulum of figure 2.12.

2.1.4.3 3D-mechanical system motion

The study of 3D-mechanical system motion is rather complicated and cannot be summed up simply. A very detailed modeling is proposed by Bos (1986).

2.1.4.4 Electrical systems

2.1.4.4.1 *Procedure*

1. Fix the circulation direction of the current.

2. For each node in the circuit with a distinct potential, place a 0-junction.

3. Insert 1-junctions between 0-junctions to indicate the potential differences and place the corresponding elements.

4. Assign the direction of the arrows which corresponds to the current's direction.

5. Choose a particular node as a voltage reference node and simplify the bond-graph where possible. The voltage reference node is suppressed as well as all the connected bonds. The following situations are equivalent:

$$\longrightarrow 1 \longrightarrow \; \equiv \; \longrightarrow \qquad \longrightarrow 0 \longrightarrow \; \equiv \; \longrightarrow$$

2.1.4.4.2 *Example*

Consider the following example:

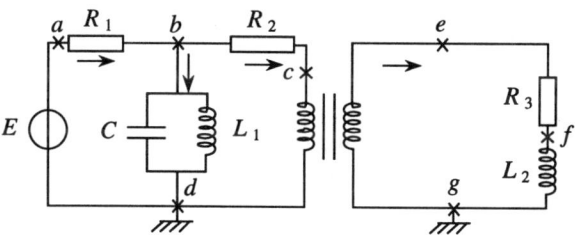

Figure 2.15: Electrical circuit.

<u>*Step 2*</u>: 0-junctions.

$$
\begin{array}{ccccc}
\underset{a}{0} & \underset{b}{0} & \underset{c}{0} & & \underset{e}{0} \\[2mm]
 & \underset{d}{0} & & & \underset{f}{0} \\[2mm]
 & & & \underset{g}{0} &
\end{array}
$$

Steps 3 and 4

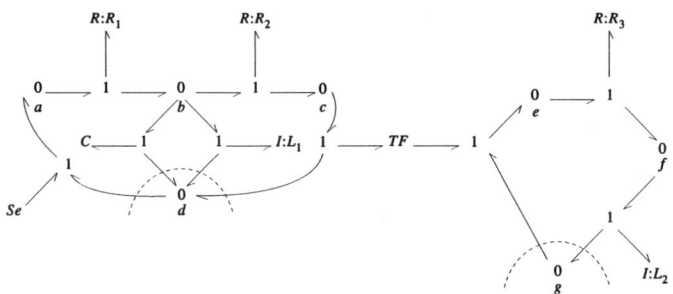

Step 5: Final bond-graph.

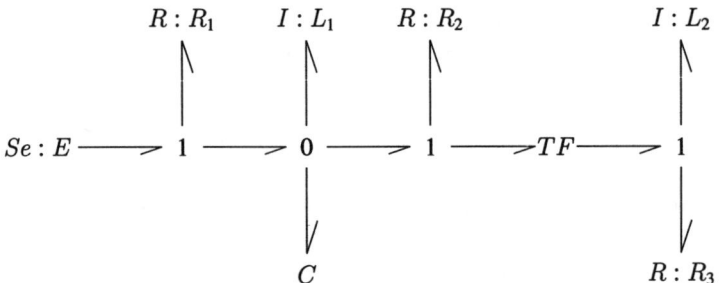

Figure 2.16: Bond-graph model associated with the electrical circuit of figure 2.15.

2.1.4.5   Hydraulic systems

The procedure is the same as in the electrical case; the 0-junctions are associated with the nodes of the circuit with a distinct pressure.

The complete bond-graph is simplified by choosing a particular pressure node which generally corresponds to the atmospheric pressure.

2.1.5   *Introduction of the causality*

A bond-graph model represents the system's architecture where the power exchanges between elements appear. This allows definition of the calculation structure by showing up the cause to effect relations in the system and then leads to a property with an obvious advantage over other graphic representations such as flow graphs.

When two subsystems A and B are connected and exchange power, $P = ef$, two situations are possible:

- A applies an "effort" e to B, which reacts by sending back to A a "flow" f;
- A sends a "flow" f to B, which responds by an "effort" e.

These two situations give rise to the following two different block-diagrams:

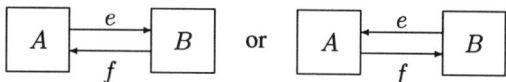

Figure 2.17: Two causality situations.

A causal stroke is introduced in order to take into account and represent these cause-to-effect relations in the bond-graph model.

By convention, the causal stroke is put near the element for which the effort is known, and thus the flow is known for the opposite element. It is represented perpendicular to the corresponding bond.

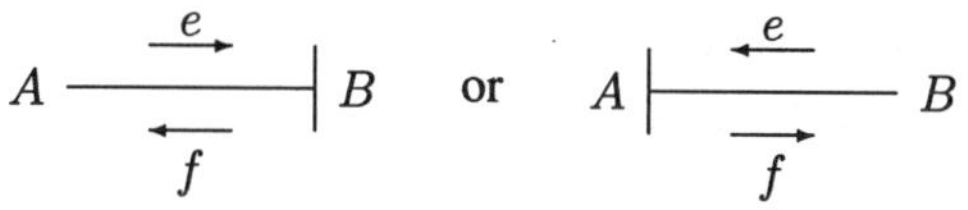

$$A \xrightarrow{\;e\;} | B \quad \text{or} \quad A | \xrightarrow{\;e\;} B$$

Figure 2.18: Causal representations corresponding to figure 2.17.

The causal stroke position is completely independent of the half-arrow direction.

In fact, if the previous example (fig. 2.17) is reconsidered, the hypothesis that the power transfer occurs from A to B leads to the orientation of the half-arrow from A to B, while two positions for the causal stroke are possible (fig. 2.19).

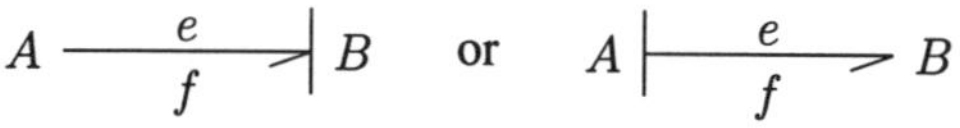

$$A \xrightarrow[f]{e} | B \quad \text{or} \quad A | \xrightarrow[f]{e} B$$

Figure 2.19: Power and causal orientations.

Figure 2.20 represents a very simple electrical diagram with the associated cause-to-effect relations in a block-diagram and causal bond-graph form.

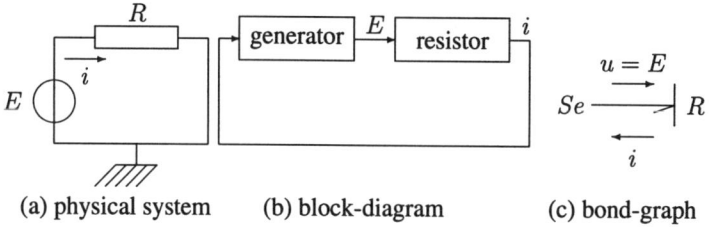

(a) physical system   (b) block-diagram   (c) bond-graph

Figure 2.20: Electrical diagram and its associated causal bond-graph.

The causality assignment is not arbitrary but follows appropriate rules presented in the following.

2.1.5.1 Required causalities

These concern the sources. The effort imposed by an effort source as well as the flow imposed by a flow source are always known variables for the system, which leads to the imposition of the causal stroke position.

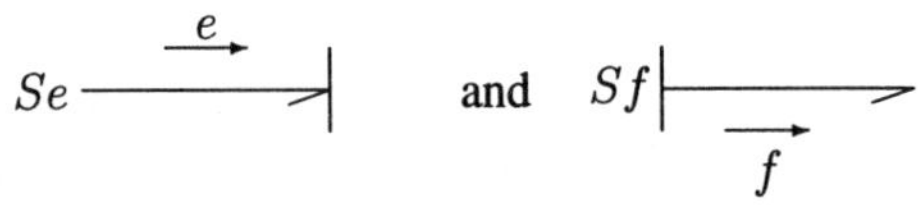

$$Se \xrightarrow{\quad e \quad} | \quad \text{and} \quad Sf | \xrightarrow{\quad\quad} $$
$$\qquad\qquad\qquad\qquad\qquad\qquad\qquad f$$

Figure 2.21: Causal assignment for the sources.

2.1.5.2   *R-, C-, I-element causalities*

2.1.5.2.1   *R-element*

In the linear case, two equivalent situations may occur:
- $e = Rf$ when f is an input for R,
- $f = (1/R)e$ when e is an input for R.

This leads to the following two causal stroke positions:

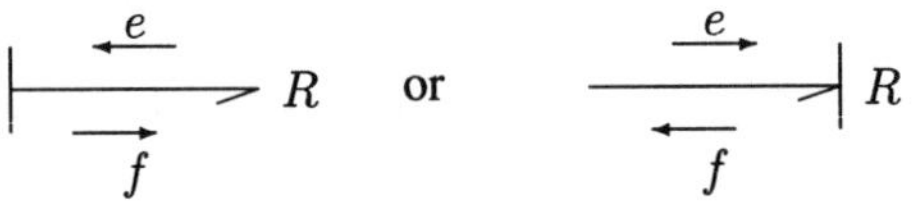

$$\vdash \xrightarrow{\quad e \quad} R \quad \text{or} \quad \xrightarrow{\quad e \quad} | R$$
$$\qquad f \qquad\qquad\qquad\qquad f$$

Figure 2.22: R-element causality.

For a linear R-element, there is no preferential causality: it adapts itself to the constraints of the context situation.

When the R-element is nonlinear (hydraulic restriction or diode, for instance), the causality is no longer arbitrary and may become an imposed constraint, as in the case of the sources.

In fact, the current-voltage characteristic of a diode is expressed by the following relations:

$$\begin{cases} \text{if} \quad u_{\text{diode}} < u_{\text{threshold}}, & \text{then} \quad i_{\text{diode}} = 0, \\ \text{if} \quad u_{\text{diode}} \geq u_{\text{threshold}}, & \text{then} \quad i_{\text{diode}} = (1/R_{\text{diode}}) \cdot (u_{\text{diode}} - u_{\text{threshold}}); \end{cases}$$

The causality assigned on the corresponding R-element is required and can be represented by:

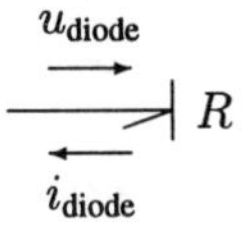

$$\xrightarrow{\quad u_{\text{diode}} \quad} | R$$
$$\xleftarrow{\qquad\qquad}$$
$$i_{\text{diode}}$$

2.1.5.2.2   *C- and I-elements*

Write the characteristic relations of the C- and I-elements in the following form:

$$e_C = \psi_C \left(\int f_C \mathrm{d}t \right), \quad \text{or} \quad e_I = \frac{\mathrm{d}}{\mathrm{d}t} \left(\psi_I^{-1}(f_I) \right), \tag{2.23}$$

supposing that f is known for the C and I and that e is a consequence. This leads us to place the causal stroke as follows:

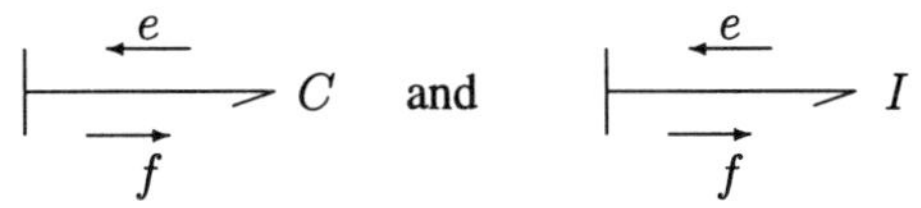

Figure 2.23: "Entering flow" causality.

To write the characteristic relations of the C- and I-elements under the following form:

$$f_C = \frac{d}{dt}\left(\psi_C^{-1}(e_C)\right), \quad \text{or} \quad f_I = \psi_I\left(\int e_I dt\right), \tag{2.24}$$

supposes on the contrary that e is known for the C- and I-elements and that f is a consequence. This leads to place the causal stroke as follows:

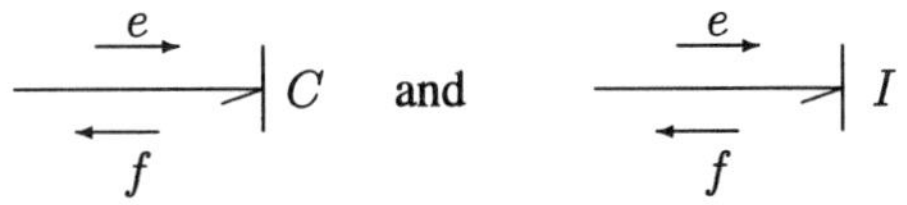

Figure 2.24: "Entering effort" causality.

For numerical and mostly physical reasons (it is easier and more robust to integrate than to derive), we will try to affect to the C- and I-elements a causality called "integral" associated to a law in the integral form. This leads to an "entering flow" causality for the C-element and an "entering effort" causality for the I-element.

Remark 3

In some cases, it is impossible to assign an integral causality to the whole set of I- and C-elements. The appearance of a derivative causality must always leads the user to wonder, and may be interpreted, in some cases, as the result of rough description of the studied phenomena.

2.1.5.3 Junction elements

2.1.5.3.1 0-junction

Let us reconsider the previous example (fig. 2.5).

The characteristic relations can be written:

$$\begin{aligned} e_1 = e_2 = e_3 = e_4 \quad &\text{(characteristic law of the 0-junction),} \\ f_1 + f_2 - f_3 - f_4 = 0 \quad &\text{(power balance).} \end{aligned} \tag{2.25}$$

This way of writing the relations is not adapted to the calculus. On the one hand, the effort which determines the value of the remaining bonds must be defined, and on the other hand, the flows to calculate have to be precised as well as the flow with a known value.

Supposing that e_2 is known, we can write:

$$e_1 = e_2, \quad e_3 = e_2, \quad e_4 = e_2,$$

which leads us to place the causal strokes as indicated in figure 2.25.

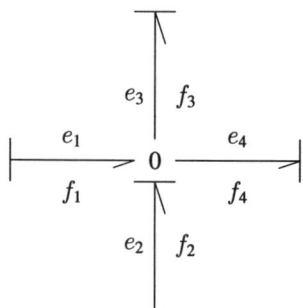

Figure 2.25: 0-junction with causality.

It gives directly the information concerning the flows and allows us to write, from (2.25):

$$f_2 = -f_1 + f_3 + f_4.$$

For a 0-junction, only one effort can impose its value on the others. We thus have the following rule:

"Only one causal stroke near a 0-junction".

2.1.5.3.2   1-junction

Let us reconsider the previous example (fig. 2.6).

The characteristic relations can be written:

$$f_1 = f_2 = f_3 = f_4 \quad \text{(characteristic law of the 1-junction)},$$
$$e_1 + e_2 - e_3 - e_4 = 0 \quad \text{(power balance).} \tag{2.26}$$

The reasoning is similar to the case of a 0-junction. A flow uniquely imposes its value on the others. Supposing it is f_3, we can write:

$$f_1 = f_3 \quad f_2 = f_3 \quad f_4 = f_3, \tag{2.27}$$

which leads to the causal bond-graph (fig. 2.26) and allows us to write:

$$e_3 = e_1 + e_2 - e_4. \tag{2.28}$$

The causality assignment rule is, in this case:

"Only one bond without a causal stroke near a 1-junction".

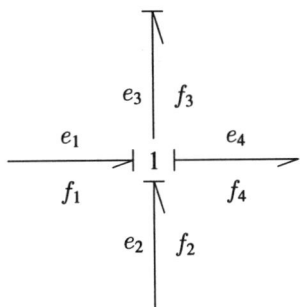

Figure 2.26: 1-junction with causality.

2.1.5.3.3 Transformer TF

Two possibilities for causal assignment are available.

The characteristic relations for a TF-element can be written, when e_2 and f_1 are known, in the following form:

$$e_1 = m\,e_2,$$
$$f_2 = m\,f_1\,;$$

(2.29)

or, when e_1 and f_2 are known, we have:

$$e_2 = (1/m)e_1,$$
$$f_1 = (1/m)f_2.$$

(2.30)

The two causal bond-graphs described in figure 2.27 are then possible.

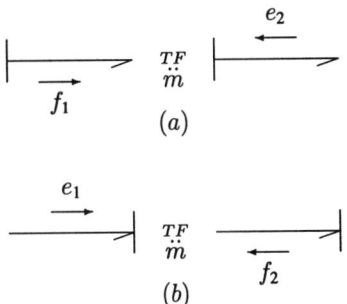

Figure 2.27: Two causal possibilities for a TF.

The causality assignment rule can be summed up as follows:

"Only one causal stroke close to a TF-element".

2.1.5.3.4   GY -element

As for a TF, two possibilities for causal assignment are available depending on the known variables for the GY.

When the flows are known, the characteristic relations can be written:

$$e_1 = rf_2,$$
$$e_2 = rf_1; \tag{2.31}$$

or, when efforts are known, these relations become:

$$f_2 = (1/r)e_1,$$
$$f_1 = (1/r)e_2. \tag{2.32}$$

The two causal bond-graphs described in figure 2.28 are then possible.

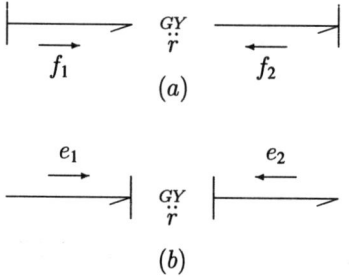

Figure 2.28: Two causal possibilities for a GY.

The causality assignment rule can be summed up as follows:

"No causal stroke or two causal strokes close to a GY-element".

2.1.5.4   Causality assignment procedure

1. Assign the required causality to the sources and consider the consequences on the the environment causality assignment.

2. Assign an integral causality to all I- and C-elements, and the required causality to nonlinear R-elements and consider the causality assignment consequences on the environment.

3. Assign causalities to junctions $0, 1, TF, GY$, using the constraints on the elements.

4. Assign causalities to the remaining linear R-elements depending on the available possibilities.

5. Search for causality conflicts. If a conflict happens, go back to step 2 and modify the causality on the I- or C-element which causes the conflict.

2.1.5.5 Examples

2.1.5.5.1 General case

Let us consider the bond-graph models (figs. 2.11 and 2.16) derived previously from the physical systems (figs. 2.10 and 2.15). The application of the causality assignment procedure allows us to obtain the following causal bond-graphs:

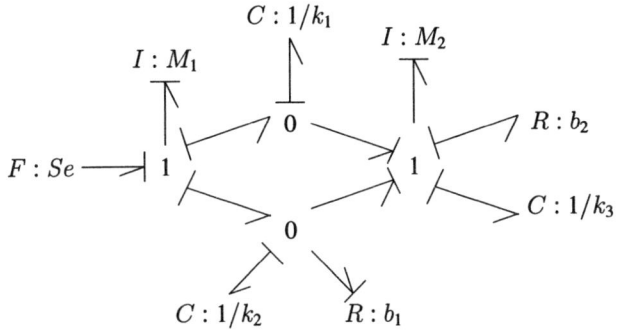

Figure 2.29: Causal bond-graph associated with the mechanical system of figure 2.10.

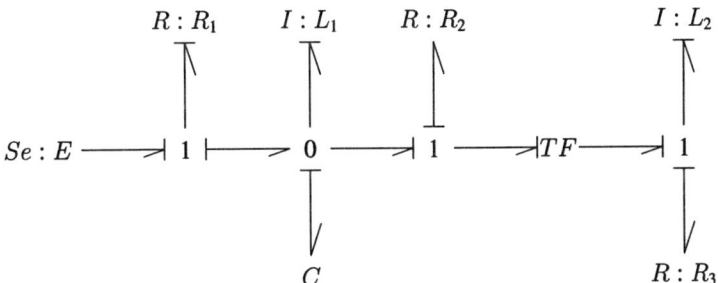

Figure 2.30: Causal bond-graph associated with the electrical system of figure 2.16.

2.1.5.5.2 Particular cases

• *Non-unique causality*

Consider the electrical system of figure 2.31, where all components are supposed to be linear, and let us assign causality to the associated bond-graph. Two situations described in figure 2.32 appear to be possible.

We notice two situations for causality assignment on bonds 2, 4, and 5, and both of them are fully valid. This is due to the particular positions of the R-elements which introduce a non-unicity of the causality assignment. This phenomenon may create numerical problems due to the appearance of implicit equations when writing the state equation (cf. Sections 2.2.4.3, and 2.2.4.3.1).

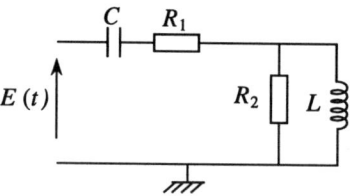

Figure 2.31: Electrical system for a bond-graph with a non-unique causality.

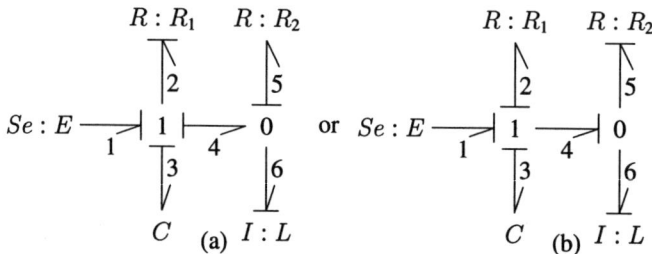

Figure 2.32: Two causality situations.

• *Derivative causality*

Consider the mechanical system of figure 2.33, and its associated bond-graph model (fig. 2.34):

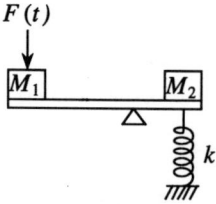

Figure 2.33: Mechanical system with a derivative causality in the associated bond-graph model.

In this example, one of the two I-elements is necessarily in derivative causality since the two inertias are linked to each other and then only one mass has an independent dynamic. The user may choose either of the two models. However, an heuristic rule implies an integral causality to I- and C-elements directly linked to an input, i.e. case (a). As for the previous case, such a situation of causality causes numerical problems. In fact, the obtained mathematical state model is an algebraic-differential system (cf. Sections 2.2.4.3 and 2.2.4.3.2).

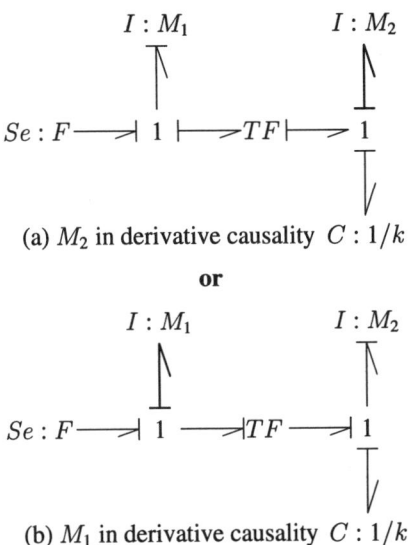

(a) M_2 in derivative causality $C : 1/k$

or

(b) M_1 in derivative causality $C : 1/k$

Figure 2.34: Two situations of derivative causalities associated with the system of figure 2.33.

2.1.6 Signal concept

2.1.6.1 Signal representation

When one of the two variables effort or flow is very weak, the transmitted power is negligible and the classical signal representation can be used:

$$\xrightarrow{\quad e \quad} \qquad \text{or} \qquad \xrightarrow[f]{\quad\quad}$$

The bond becomes an information bond instead of a power bond.

This type of bond shows up in the bond-graph model of detectors and sensors (supposed ideal), the calculus operators such as integrators, summers or comparators, and any compensation network which delivers a control signal result of a calculus carried out using feedback information.

$$\longrightarrow\!\!| D : D_e \qquad \text{and} \qquad \longmapsto D : D_f$$

The effort and flow detectors are represented, with a joint use of the signal representation and the causality assignment.

The bond-graph is not a graphic tool disconnected from the other graphical tools such as flow graphs and even Petri nets. On the contrary, it intervenes to represent in a more complete way the physical phenomena which appear in a system, and can perfectly be included as a knowledge model, when simulating, in a classical feedback control structure such as the one presented in figure 2.35.

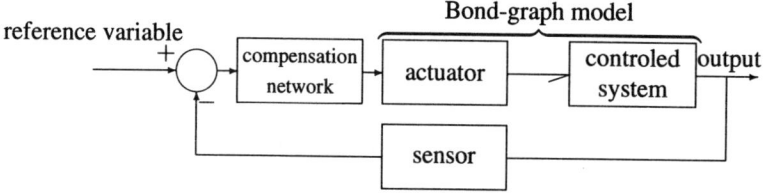

Figure 2.35: Bond-graph model in a block-diagram.

2.1.6.2   Controled source

When the source is not independent, but depends on the context, the controled source concept may be used. This formalism is very useful to model closed-loop structures.

Different types of controled sources can be considered:

- Effort source controled by effort with $e_2 = k(e_1)$ (fig. 2.36 a);
- Effort source controled by flow with $e_2 = k(f_1)$ (fig. 2.36 b);
- Flow source controled by effort with $f_2 = k(e_1)$ (fig. 2.36 c);
- Flow source controled by flow with $f_2 = k(f_1)$ (fig. 2.36 d).

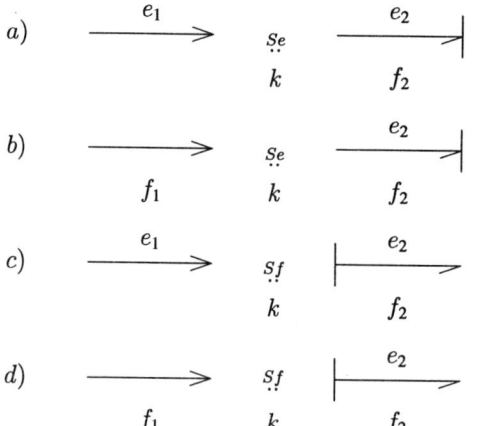

Figure 2.36: Controled sources.

2.1.7   Conclusion

Bond-graph elements presented previously are simple elements (called 1-port since they gather a couple of variables) which allow the representation of most physical phenomena. Table 2.2 includes the basic bond-graph elements and their properties.

Nevertheless, when considering complex systems, the different laws involved are no more scalar and multiport elements characterized by vectorial laws (Karnopp, 1979; Karnopp et al, 1990; Thoma, 1990) have to be used. An example of the use of these elements is presented in section 2.1.3.2.

2.2 Bond-graph causal properties

The bond-graph model of a dynamic system is situated half way between the physical system and the associated mathematical models (state equation, transfer function or matrix). The bond-graph shows not only the system's architecture but also its causal organization due to its ability to exhibit the cause to effect relations between elements.

This property is fundamental since it supposes the use of a systematic and well organized approach for the writing of the characteristic relations of the system evolution, and permits the combination of differential and algebraic equations.

However, some other possibilities of great interest linked to the graphic aspect of a bond-graph and to its causal structure are available for the user. They are obtained by running through the bond-graph model following privileged paths, called causal paths, independently from the power direction of the bonds.

2.2.1 Constitutive laws for R-, C-, I-elements

The constitutive laws for R-, C-, I-elements correspond to the laws which link the input variable of the elements to their output variable. Thus,

- For a C-element in integral causality, the elementary relation can be written:

$$\begin{array}{c} e_C \\ \longleftarrow \\ \vdash\!\!\!\longrightarrow C \\ \longrightarrow \\ f_C \end{array}$$

$e_c = \psi_c(\int f_c \mathrm{d}t)$ or $e_c = \psi_c(q_c)$. In the linear case and using the Laplace operator s, the transmittance for a C-element can be defined by:

$$\frac{E_c(s)}{F_c(s)} = \frac{1}{Cs}.$$

- For a C-element in derivative causality, the elementary relation can be written:

$$\begin{array}{c} e_C \\ \longrightarrow \\ \longrightarrow\!\!\!\vdash C \\ \longleftarrow \\ f_C \end{array}$$

$$f_c = \frac{\mathrm{d}}{\mathrm{d}t}\left(\psi_c^{-1}(e_c)\right) \quad \text{or} \quad q_c = \psi_c^{-1}(e_c) \quad \text{in the general case,}$$

BASIC BOND-GRAPH ELEMENTS AND THEIR PROPERTIES

Bond-graph elements	Symbols	Generic law	Physical examples	Causality
Power variables	effort e flow f		force, torque, voltage, pressure... velocity, ang. vel., current, vol. rate...	
Energy variables	momentum p displacement q	$p = \int e\,dt$ $q = \int f\,dt$	impulse, flow (self),... displacement, charge, volume...	
Active elements (sources)	$Se \xrightarrow{\;e\;}{f}$ $Sf \xrightarrow{\;e\;}{f}$	e independent of f f independent of e	gravity, voltage generator,... current generator, pump,...	$Se \xrightarrow{\;e\;}$ e imposed $Sf \vdash\!\!\xrightarrow{}{f}$ f imposed
Passive 1-port elements	$\xrightarrow{\;e\;}{f}\, R$	$\Phi_R(e,f) = 0$	damper, resistor, hydraulic restriction, friction,...	$\vdash\!\!\rightarrow\!R$ $\;e = F_R(f)\;(e = Rf)$ $\rightarrow\!\!\dashv R$ $\;f = F_R^{-1}(e)\,(f = e/R)$
	$\xrightarrow{\;e\;}{f = \dot{q}}\, C$	$\Phi_C(e,q) = 0$	spring, capacitor, tank, elasticity,...	$\vdash\!\!\rightarrow\!C$ $\;e = F_C(q)(e = q/C)$ $\rightarrow\!\!\dashv C$ $\;q = F_C^{-1}(e)(q = Ce)$
	$\xrightarrow{e = \dot{p}}{f}\, I$	$\Phi_I(p,f) = 0$	mass, inertia, inductance,...	$\rightarrow\!\!\dashv I$ $\;f = F_I(p)\;(f = p/I)$ $\vdash\!\!\rightarrow\!I$ $\;p = F_I^{-1}(f)(p = If)$

SUMMARY OF THE BASIC BOND-GRAPH ELEMENTS (CONTINUED).

Bond-graph elements	Symbols	Generic law	Physical examples	Causality
		$e_1 = \ldots = e_n$ $\sum_{i=1}^{n} a_i f_i = 0 \ (a_i = \pm 1)$	series connexion in mechanics parallel connexion in electrical domain	ex: $e_1 = e_2, \ldots e_n = e_2, \quad f_2 = -\sum_{i=1, i \neq 2}^{n} a_i f_i$
Junction elements		$f_1 = \ldots = f_n$ $\sum_{i=1}^{n} a_i e_i = 0 \ (a_i = \pm 1)$	parallel connexion in mechanics series connexion in electrical domain	ex: $f_1 = f_2 \ldots f_n = f_2, \quad e_2 = -\sum_{i=1, i \neq 2}^{n} a_i e_i$
(power conservative)	$e_1 \xrightarrow{\ } TF \xrightarrow{\ } e_2$ $f_1 \quad \overset{m}{} \quad f_2$	$e_1 = m e_2$ $f_2 = m f_1$	lever, gears, pulleys electrical transformer	$\xrightarrow{\ } TF \xleftarrow{\ } \ \begin{cases} e_1 = m e_2 \\ f_2 = m f_1 \end{cases}$ $\xleftarrow{\ } TF \xrightarrow{\ } \ \begin{cases} e_2 = 1/m\, e_1 \\ f_1 = 1/m\, f_2 \end{cases}$
	$e_1 \xrightarrow{\ } GY \xrightarrow{\ } e_2$ $f_1 \quad \overset{r}{} \quad f_2$	$e_1 = r f_2$ $e_2 = r f_1$	Hall effect sensor, gyroscope, DC-motor,....	$\xrightarrow{\ } GY \xleftarrow{\ } \ \begin{cases} e_1 = r f_2 \\ f_2 = r f_1 \end{cases}$ $\xleftarrow{\ } GY \xrightarrow{\ } \ \begin{cases} f_2 = 1/r\, e_1 \\ f_1 = 1/r\, e_2 \end{cases}$

TABLE 2.2: SUMMARY OF THE BASIC BOND-GRAPH ELEMENTS.

and leads, in the linear case to the following transmittance:

$$\frac{F_c(s)}{E_c(s)} = Cs.$$

Likewise, the characteristic relation for an I-element depends on its causality:

- I in integral causality:

$$f_I = \psi_I \left(\int e_I \mathrm{dt} \right) \quad \text{or} \quad f_I = \psi_I(p_I)$$

$$\text{or} \quad \frac{F_I(s)}{E_I(s)} = \frac{1}{Is} \, ;$$

- I in derivative causality:

$$e_I = \frac{\mathrm{d}}{\mathrm{dt}} \left(\psi_I^{-1}(f_I) \right) \quad \text{or} \quad p_I = \psi_I^{-1}(f_I)$$

$$\text{or} \quad \frac{E_I(s)}{F_I(s)} = Is.$$

An R-element has two different situations for the causality assignment which lead to two types of laws:

$$e_R = \psi_R(f_R) \quad \text{or} \quad \frac{e_R}{f_R} = R \, ;$$

$$f_R = \psi_R^{-1}(e_R) \quad \text{or} \quad \frac{f_R}{e_R} = \frac{1}{R}$$

2.2.2 Construction of a block-diagram associated with a bond-graph

Until now, only some software programs allowed bond-graph models for simulation purposes. The more common way to enter models was under block-diagrams form, which can be directly deduced from a bond-graph model.

Consider the electrical example (fig. 2.37) and its associated bond-graph model (fig. 2.38). The block-diagram can be derived from the bond-graph just by writing successively the structural laws at the junctions and the constitutive laws for the elements.

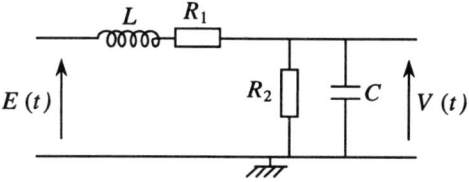

Figure 2.37: Electrical circuit.

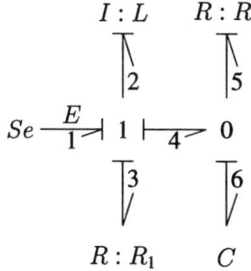

Figure 2.38: Bond-graph model associated with the electrical circuit (fig. 2.37).

- Structural laws

  - 1-junction:

    $f_1 = f_2 = f_3 = f_4$ (characteristic law of a 1-junction),

    $e_1 - e_2 - e_3 - e_4 = 0$ (power balance),

    and owing to the causality, we have:

    $$f_1 = f_2, \quad f_3 = f_2, \quad f_4 = f_2, \qquad e_2 = e_1 - e_3 - e_4. \qquad (2.33)$$

  - 0-junction:

    $e_4 = e_5 = e_6$ (characteristic law of a 0-junction),

    $f_4 - f_5 - f_6 = 0$ (power balance),

    and owing to the causality, we have:

    $$e_4 = e_6, \quad e_5 = e_6, \qquad f_6 = f_4 - f_5. \qquad (2.34)$$

- Constitutive laws for the elements.

$$f_2 = \psi_L \left(\int e_2 dt \right),$$

$$e_6 = \psi_C \left(\int f_6 dt \right),$$

$$e_3 = \psi_{R_1}(f_3),$$

$$f_5 = \psi_{R_2}^{-1}(e_5). \qquad (2.35)$$

Figure 2.39 is obtained as follows: show in two distinct lines, associated respectively with efforts and flows, a summation and a connexion nodes for each junction of the bond-graph to represent the weighted sum for a type of variable and the equality for the other, then just write out relations (2.33), (2.34), (2.35).

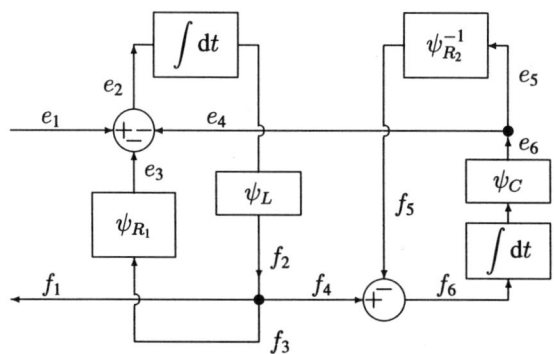

Figure 2.39: Block-diagram associated with the bond-graph model (fig. 2.38).

From the obtained structure, the corresponding simplified block-diagram can be derived (fig. 2.40), by running through the graph from the input signal $e_1 = E$, and following just one variable at a time.

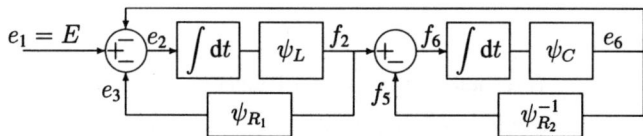

Figure 2.40: Simplified block-diagram associated with the electrical circuit (fig. 2.37).

Remark 4

  − *Some more complex cases may occur, for example, when a GY-element appears in the loops. For more details see Brown (1972).*
  − *This method can be applied, without any problem, to the multivariable case.*

2.2.3   *Causal path - Causal loop*

2.2.3.1   Definitions

  1. A causal path in a bond-graph junction structure is the alternation of bonds and basic elements, called "nodes" such as:

     (a) for the acausal bond-graph, the sequence forms a simple chain,

     (b) all nodes in the sequence have complete and correct causality,

(c) two bonds from a causal path have at the same node opposite causal orientations.

2. A causal path is **simple** if it can be run through while always following the same variable. In the same sequence of bonds and nodes there are therefore two paths by following either effort or flow as illustrated in figure 2.41.

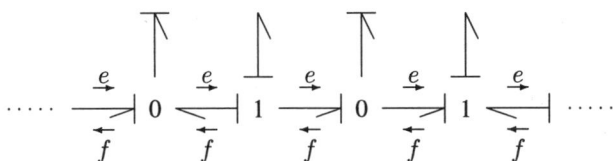

Figure 2.41: Simple causal paths in a bond-graph model.

3. A causal path is **mixed** if a change of variable is needed while running through the graph. This is the case when a GY is present (fig. 2.42 (a)), and the causal path is called **mixed direct**, or when an R-, C-, I-element has to be gone through and the causal path is called **mixed indirect** (fig. 2.42 (b)).

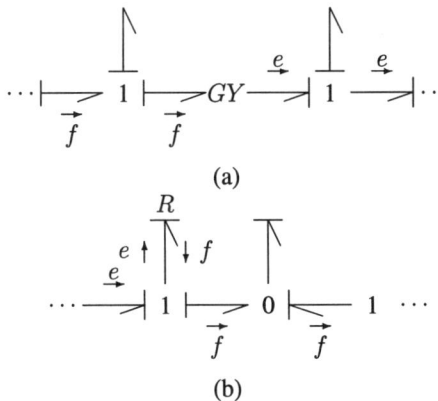

Figure 2.42: Causal path mixed direct (a) and indirect (b).

4. Two elements P_1 and P_2, belonging to the set $\{R, C, I, S_e, S_f, D_e, D_f\}$ are **causally connected** if the input variable of one of then is influenced by the output variable of the other as indicated in figure 2.43.

Thus, the output of $P_1 : e_{p_1}$ (resp. of $P_2 : f_{p_2}$), influences the input of $P_2 : e_{p_2}$ (resp. of $P_1 : f_{p_1}$).

5. A **forward path** is a causal path between an input and a detector.

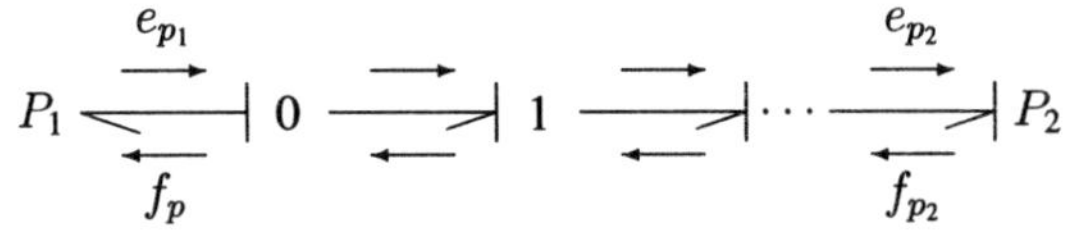

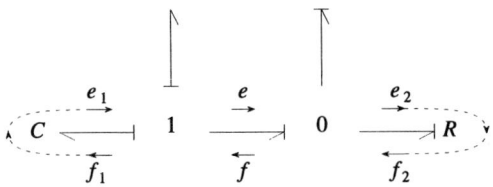

Figure 2.43: Causal connexion between two elements.

6. A **causal loop** (fig. 2.44) is a closed causal path starting from the output of an R-, C- or I-element and ending at the input of this element without running through the same bond while following the same variable more than once.

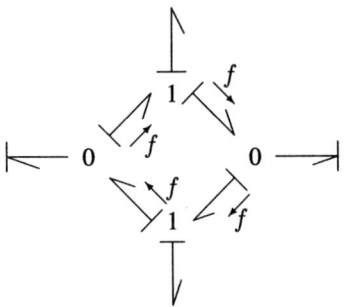

Figure 2.44: Causal loop between C and R.

7. A **causality loop** is a succession of junctions and bonds forming a cycle and for which the causality is oriented in the same direction for all bonds (fig. 2.45), (except in the presence of a GY).

Figure 2.45: Causality loop.

2.2.3.2 Characteristic function of a causal path

The characteristic function of a causal path is defined as the function which links the output variable of the element corresponding to the origin of the path, to the input variable of the element forming the end of the path. Thus, the causal path which links the source to the I-element, following bonds $1 - 2 - 3 - 4 - 5$ in Figure 2.46 gives a relationship between the effort variables e_1 and e_5, in the form $e_5 = \text{function}(e_1)$. This relation may be linear (called causal path gain) or nonlinear.

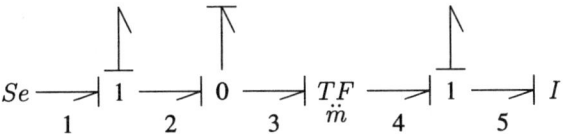

Figure 2.46: Causal path between Se and I.

2.2.3.2.1 Simple or mixed direct causal path

The characteristic function for a simple or mixed direct causal path, indexed i, is calculated by:

$$T_i = (-1)^{n_0+n_1} \cdot \prod_{j,k} \left(m_j \text{ or } \frac{1}{m_j} \right) \cdot \left(r_k \text{ or } \frac{1}{r_k} \right), \tag{2.36}$$

where:

• n_0 and n_1 represent the total number of changes of orientations for the bonds, respectively at the 0-junctions while following the flow variable, and at the 1-junctions while following the effort variable (figure 2.47).

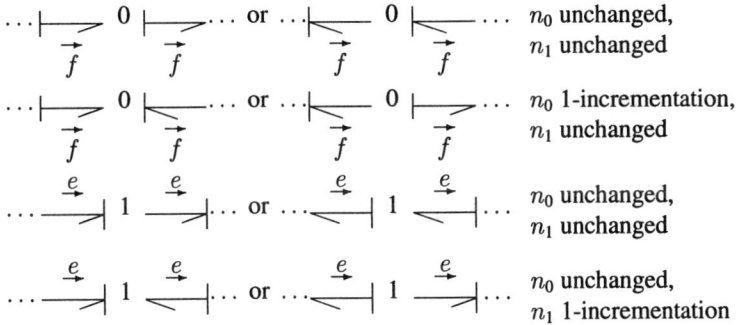

Figure 2.47: Calculus of the exponents n_0 and n_1.

• m_j, $1/m_j$ and r_k, $1/r_k$ are the moduli of the TF-elements (or MTF) and GY (or MGY) which intervene in the causal path, depending on their causalities.

Consider the example presented in figure 2.48.

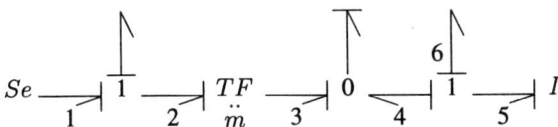

Figure 2.48: Example of a simple causal path.

The causal path linking the source Se to the I-element is run through, following the effort along bonds $1-2-3-4-5$, and n_0 is immediately fixed equal to zero. The exponent n_1 is equal to 1 here, because both bonds 4 and 5 are going out of the 1-junction (fig. 2.47).

The causality applied to the TF-element gives the term $1/m$, which leads to the following final value:

$$T_1 = \frac{e_5}{e_1} = (-1)\left(\frac{1}{m}\right).\tag{2.37}$$

The characteristic function of a causal path may be nonlinear when the moduli m_j and r_k of the TF and GY are non-constant. This is the case in particular for mechanical systems in plane or space motion for which the junction structure represents the kinematic relations and then is expressed in terms of trigonometric functions of the angles between bodies (section 2.1.1.4.2).

Indeed, from the bond-graph model of an elastic pendulum, presented in figure 2.14, completed by the causality assignment (fig. 2.49), it is possible to obtain non-constant causal path gains. This is the case, for example, for the causal path going from the I-element (bond 3) to the R-element, and running through bonds $3-11-10-4$:

$$T_1 = (-1)^{n_0+n_1}\left(\frac{1}{r\cos\theta}\right) \quad \text{with} \quad n_0 = 0 \quad n_1 = 0.\tag{2.38}$$

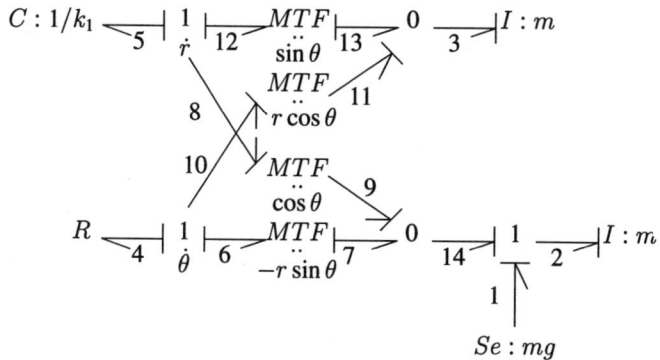

Figure 2.49: Complete bond-graph model for the elastic pendulum of figure 2.12.

2.2.3.2.2 Indirect causal path

When the causal path is indirect, going through R-, C-, or I-elements, the characteristic relations of these elements must be taken into consideration for the calculus of the causal path gain.

When the elements and the structure are linear, the expression obtained by equation (2.36) has to be multiplied by the transmittance of the traversed elements.

When the elements are nonlinear, the characteristic relation of the elements must be expressed. In fact, consider the causal path issued from figure 2.49, linking the source Se and the C-element: $1/k_1$, going through the I-element and represented in figure 2.50.

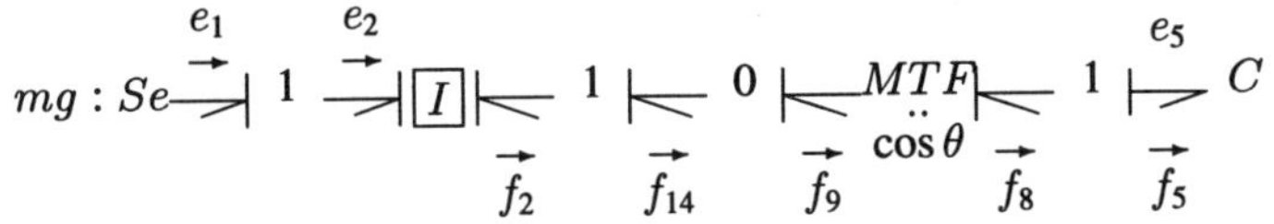

$$e_1 \qquad e_2 \qquad\qquad\qquad\qquad\qquad\qquad e_5$$

$$mg : Se \xrightarrow{\quad} 1 \xrightarrow{\quad} \boxed{I} \longleftarrow 1 \longleftarrow 0 \longleftarrow MTF \longleftarrow 1 \xrightarrow{\quad} C$$
$$\qquad\qquad\qquad\qquad\qquad\qquad \overset{\cdots}{\cos\theta}$$
$$\qquad\qquad f_2 \qquad f_{14} \qquad f_9 \qquad f_8 \qquad f_5$$

Figure 2.50: A causal path in the bond-graph model of figure 2.49.

If I is nonlinear, characterized by the relation $f_2 = \psi_I \left(\int e_2 dt \right)$, then the characteristic function of the $Se - C$ path should be written:

$$f_5 = \left(\frac{1}{\cos\theta} \right) \psi_I \left(\int e_1 dt \right). \tag{2.39}$$

If the desired function is that which links efforts e_5 and e_1, the characteristic relation of the C-element must then be taken into consideration, that is, in the general case, $e_5 = \psi_C \left(\int f_5 dt \right)$, and we have:

$$e_5 = \psi_C \left\{ \int \left[\frac{1}{\cos\theta} \psi_I \left(\int e_1 dt \right) \right] dt \right\}. \tag{2.40}$$

2.2.3.3 Characteristic function of a causal loop

The same distinction as for the causal path discussion, has to be made between the linear and nonlinear cases.

2.2.3.3.1 *Case when R-, C-, I- and TF-, GY-elements are linear*

When two elements from the set (R, C, I) are causally connected and compose a causal loop indexed i, the gain of the loop B_i is obtained by the following relation:

$$B_i = (-1)^{n_0 + n_1} . \prod_{j,k} (m_j \text{ or } \frac{1}{m_j})^2 . (r_k \text{ or } \frac{1}{r_k})^2 . \prod_g, \tag{2.41}$$

where n_0 and n_1 are defined as previously and $\prod_g$ designs the product of gains of the elements composing the loop.

All terms of expression (2.41) are constant, the product is commutative, and there is no need to specify the starting and ending points of the loop. Let us reconsider, as an example, the bond-graph model (fig. 2.38) and its associated electrical circuit (fig. 2.37).

Three causal loops appear in the bond-graph:
• Between R_1 and L:

$$R : R_1 \longleftarrow\!\!| \; 1 \; \underset{2}{\longrightarrow}\!| \; I : L$$
$$ {}_3$$

the gain of the loop is $B_1 = -R_1/Ls$, in accordance with the previous formula (2.41).
The sign is due to a change of the half-arrow directions at a 1-junction when following the
effort (path from 3 to 2), that is $n_0 = 0$ and $n_1 = 1$.
• Between C and R_2:

$$C \; \underset{6}{\longleftarrow}\!| \; 0 \; \underset{5}{\longrightarrow}\!| R : R_2$$

The gain of this causal loop is:

$$B_2 = -1/R_2Cs, \qquad (n_0 = 1, n_1 = 0).$$

• Between L and C:

$$I : L \; |\!\!\longleftarrow \; \underset{2}{1} \; |\!\longrightarrow \; \underset{4}{0} \; |\!\longrightarrow \; \underset{6}{C}$$

The gain of this causal loop is:

$$B_3 = -1/LCs^2, \qquad (n_0 = 0, n_1 = 1).$$

2.2.3.3.2 Case when R-, C-, I- or TF-, GY-elements are nonlinear

In this case, the causal loop gain concept cannot be used and must be replaced by the
causal loop characteristic function, defined by extension of equation (2.41).

Consider the partial bond-graph (fig. 2.51) which shows a causal connexion between
R- and C-elements, supposed nonlinear, and the associated block-diagram (fig. 2.52).
The characteristic loop function depends on the element considered as starting and ending
point. Suppose that C is this particular point, then the characteristic loop function is the
following law, expressed for the opened loop and linking f_1 to itself:

$$-\left(\frac{1}{m}\right) \cdot \psi_R^{-1} \left[\frac{1}{m} \psi_C \left(\int dt \right) \right].$$

By way of checking, it is obvious, when the characteristic laws for the R- and C-elements
are linear, that the previous function is expressed as $-1/m \times 1/R \times (1/m)(1/Cs)$, which
corresponds exactly to the result obtained directly by applying expression (2.41).

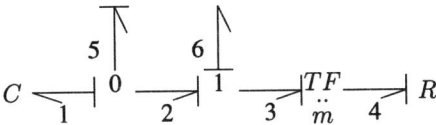

Figure 2.51: Causal loop between C and R.

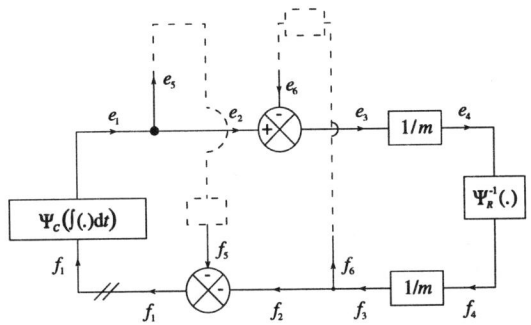

Figure 2.52: Block-diagram associated with the causal loop of figure 2.51.

2.2.4 Construction of the state equation associated with a bond-graph model

Many numerical simulations need the use of state space representation. The bond-graph model leads to a state equation based on the definition of a particular state vector.

2.2.4.1 State vector

Definition 2.2.4.1 *The state vector x is composed of the energy variables p and q associated with the I- and C-elements.*

$$x = \begin{bmatrix} p_I \\ q_C \end{bmatrix}.$$

Properties
- The state vector does not appear in the bond-graph; only its derivative does:

$$\dot{x} = \begin{bmatrix} e_I \\ f_C \end{bmatrix}.$$

- When all I- and C-elements are in integral causality, the dimension of x gives the model's order.
- If, among the n I- and C-elements, n_d are in derivative causality, then the state vector is decomposed into x^i whose dimension is n_i, and x^d, whose dimension is n_d (i for "integral" and d for "derivative").

2.2.4.2   Systematic method

To obtain the state equation in the form:

$$\dot{x} = f(x, u),$$
$$y = h(x),$$
(2.42)

we can proceed as follows:

  - write the structure laws associated with the junctions by taking into consideration the causality;

  - write the constitutive laws for the elements;

  - combine the different previous relations so as to obtain the derivative state variables and then the state equation.

Let us reconsider the electrical example (fig. 2.37) and its associated bond-graph model with the derivative state variables (fig. 2.53).

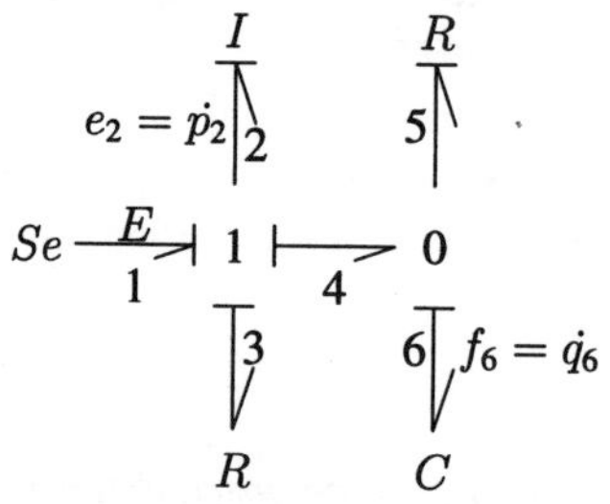

Figure 2.53: Bond-graph model with state variables.

The dimension of the state vector is 2, as well as the order of the system (I and C in integral causality).

• Structural laws:

  - 1-junction:

$$f_1 = f_2, \quad f_3 = f_2, \quad f_4 = f_2,$$
$$e_2 = e_1 - e_3 - e_4.$$
(2.43)

  - 0-junction:

$$e_4 = e_6, \quad e_5 = e_6,$$
$$f_6 = f_4 - f_5.$$
(2.44)

- Constitutive laws for the elements:

$$f_2 = \psi_L \left(\int e_2 dt \right) = \psi_L(p_2),$$

$$e_6 = \psi_C \left(\int f_6 dt \right) = \psi_C(q_6),$$

$$e_3 = \psi_{R_1}(f_3),$$

$$f_5 = \psi_{R_2}^{-1}(e_5).$$

(2.45)

Thus we have:

$$\dot{p}_2 = e_2,$$

and, from (2.43) and (2.45):

$$\dot{p}_2 = E - \psi_{R_1}(f_3) - e_6,$$

then:

$$\dot{p}_2 = E - \psi_{R_1}(\psi_L(p_2)) - \psi_C(q_6).$$

Likewise:

$$\dot{q}_6 = f_6, \qquad \text{then} \qquad \dot{q}_6 = f_2 - \psi_{R_2}^{-1}(e_5),$$

$$\dot{q}_6 = \psi_L(p_2) - \psi_{R_2}^{-1}(\psi_C(q_6)).$$

The state equation can be written as:

$$\begin{pmatrix} \dot{p}_2 \\ \dot{q}_6 \end{pmatrix} = \begin{bmatrix} -\psi_{R_1}(\psi_L(p_2)) - \psi_C(q_6) \\ \psi_L(p_2) - \psi_{R_2}^{-1}(\psi_C(q_6)) \end{bmatrix} + \begin{bmatrix} 1 \\ 0 \end{bmatrix} E.$$

(2.46)

For the output equation, if we choose $y = V(t)$, the voltage in the C-element, is:

$$y = V = e_6 = \psi_C(q_6).$$

(2.47)

Remark 5
Equation (2.46) can be written:

$$\begin{pmatrix} \dot{p}_2 \\ \dot{q}_6 \end{pmatrix} = \begin{bmatrix} -\psi_{R_1} o \psi_L & -\psi_C \\ \psi_L & -\psi_{R_2}^{-1} o \psi_C \end{bmatrix} \begin{pmatrix} p_2 \\ q_6 \end{pmatrix} + \begin{pmatrix} 1 \\ 0 \end{pmatrix} E,$$

(2.48)

that is $\dot{x} = f(x) + BE$.

It is interesting to note that terms appearing in equation (2.48) can be obtained directly from the characteristic causal loop and path functions defined previously.

Indeed, let us reconsider the block-diagram (fig. 2.40) associated with the bond-graph (fig. 2.53), and modified to show up the state variables (fig. 2.54). It clearly appears that the terms $-\psi_{R_1} o \psi_L$ and $-\psi_{R_2}^{-1} o \psi_C$ correspond to the relations linking both $\dot{p}_2$ to p_2, and $\dot{q}_6$ to q_6, obtained by combining the characteristic laws for R_1 and I, or R_2 and C, and linked to the characteristic causal loop functions between R_1 and I, or R_2 and C, without the integral $\int dt$ part.

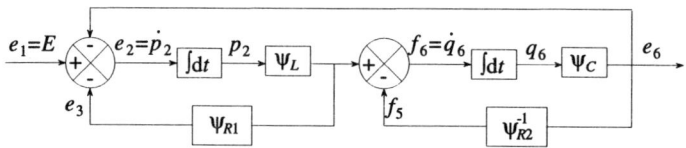

Figure 2.54: Block-diagram associated with the bond-graph model of figure 2.53.

The terms $-\psi_C$ and ψ_L can be interpreted respectively as:

(characteristic function of the causal path starting from C and going to I) $\times$ constitutive law of the C-element,

and:

(characteristic function of the causal path starting from I and going to C) $\times$ constitutive law of the I-element.

The path characteristic functions are reduced to signs when the paths are direct without any TF-element or GY-element.

Likewise, the control matrix $B = \begin{pmatrix} 1 \\ 0 \end{pmatrix}$ contains the causal path gains between the input $Se : E$ and each dynamic element.

These remarks enable an informed user to derive directly the state equation associated with a bond-graph model just by picking out the causal paths and loops.

2.2.4.3 Particular cases

2.2.4.3.1 *Case when the causality is not unique*

Let us reconsider the example presented in section 2.1.5.5.2 (fig. 2.32), completed with the derivations of the state variables (fig. 2.55).

The causality appearing in figure 2.55 is not unique, because of the existence of a causal loop between the two R-elements of bonds 2 and 5.

Consider case (a) and let us write the relation which links $\dot{p}_6$ to the state variables:

$$\dot{p}_6 = e_6 = e_5 = \psi_{R_2}(f_5),$$

that is:

$$\dot{p}_6 = \psi_{R_2} \left[\psi_{R_1}^{-1}(E - \psi_C(q_3) - \dot{p}_6) - \psi_L(p_6) \right].$$

This implicit equation leads to numerical problems which may be solved using the bond-graph.

In fact, to avoid the coupling between the two R-elements and the ambiguity in the causal assignment, the causal loop must be "broken" by the introduction of an other element, for instance. We can notice that, if the physical circuit is modified to show an inductance (the value of which is weak) for electrical line (fig. 2.56), the ambiguity in the choice of the

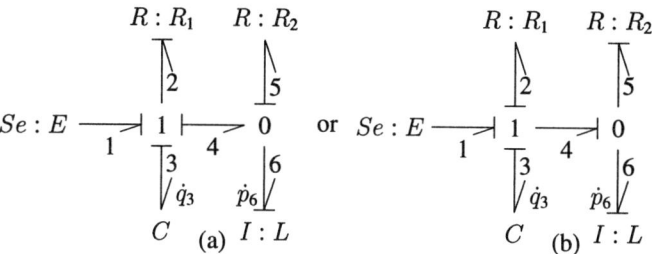

Figure 2.55: Causality not unique.

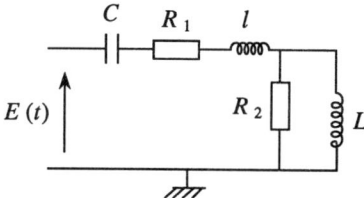

Figure 2.56: Electrical system with inductance of the wire.

causality for the 1-junction disappears (fig. 2.57). This implies an increase of the state vector's dimension because of the new variable associated with l, and can cause problems of numerical stiffness because of the very rapid mode introduced.

It is up to the user to choose between the two uncomfortable situations: either the implicit equation or the stiff equation!

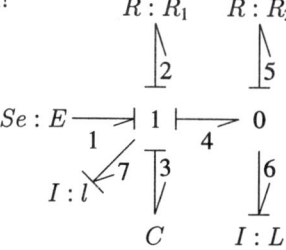

Figure 2.57: Transformed bond-graph.

2.2.4.3.2 *Case when the I- or C-elements are in derivative causality*

Consider the physical system and its associated bond-graph model (fig. 2.58, (a) and (b)).

The causality restrictions on the 0-junction impose a C-element in a derivative causality, which we have arbitrarily chosen $C : C_1$. The order of the state model is 1; taking into account the choice for the causality assignment, only q_5 has an independent dynamic. This can be explained physically by the existence of an equivalent C-element which combines C_1 and C_2.

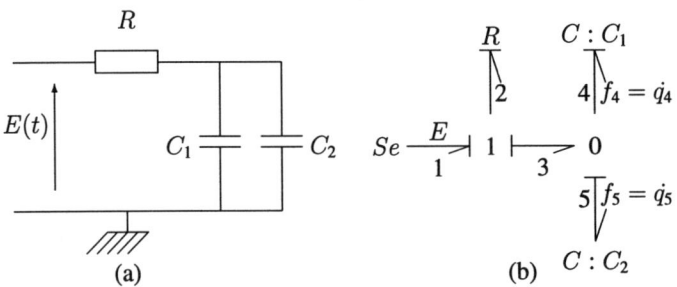

Figure 2.58: Electrical circuit and the associated bond-graph.

The state equation can be written:

$$\dot{q}_5 = f_5 = f_3 - f_4 = \psi_R^{-1}\left(E - \psi_{C_2}(q_5)\right) - \dot{q}_4, \tag{2.49}$$

with:

$$\dot{q}_4 = \frac{d}{dt}\left(\psi_{C_1}^{-1}(e_4)\right) = \frac{d}{dt}\left(\psi_{C_1}^{-1}\left(\psi_{C_2}(q_5)\right)\right). \tag{2.50}$$

If the nonlinear laws for C_1 can be derived, the obtained model is algebraic-differential. This type of model involves also numerical problems when the software is not adapted. To eliminate this difficulty, it is possible to add elements, as in the case of causal loops between R-elements, while trying to justify physically the choice and the position of added element.

2.2.4.4   Matrix method

This method consists in regrouping, under different vector forms, variables which intervene in the bond-graph model.

A bond-graph can be represented as follows:

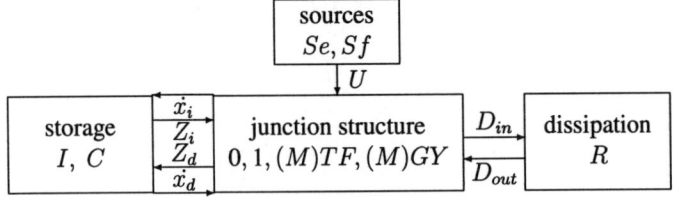

Figure 2.59: Vector representation of a bond-graph.

Let us note D_{in} and D_{out} the vectors composed of efforts and flows respectively going inside and outside the R-elements, x^i and x^d the state variables p and q associated with I- and C-elements respectively in integral and derivative causality, and Z^i and Z^d the "complementary state vectors" associated with x^i and x^d, and composed of efforts and flows.

All these vectors satisfy the following relations, formed from the constitutive laws of the elements:

$$D_{out} = LD_{in}, \quad \text{with} \quad L = \begin{bmatrix} R & 0 \\ 0 & R^{-1} \end{bmatrix}, \quad \text{or} \quad D_{out} = L(D_{in}),$$

$$Z^i = F^i x^i, \quad \text{with} \quad F^i = \begin{bmatrix} (I^i)^{-1} & 0 \\ 0 & (C^i)^{-1} \end{bmatrix}, \quad \text{or} \quad Z^i = F^i(x^i), \qquad (2.51)$$

$$Z^d = F^d x^d, \quad \text{with} \quad F^d = \begin{bmatrix} (I^d)^{-1} & 0 \\ 0 & (C^d)^{-1} \end{bmatrix}, \quad \text{or} \quad Z^d = F^d(x^d),$$

depending on whether the R, I and C are linear or not.

In the linear case (and when R-, C-, and I-elements are simple), the L, F^i and F^d matrices are diagonal.

The junction-structure is characterized by the S-matrix called "junction structure matrix", constructed using the following relation:

$$\begin{bmatrix} \dot{x}^i \\ Z^d \\ D_{in} \end{bmatrix} = \begin{bmatrix} S_{11} & S_{12} & S_{13} & S_{14} \\ S_{21} & S_{22} & S_{23} & S_{24} \\ S_{31} & S_{32} & S_{33} & S_{34} \end{bmatrix} \begin{bmatrix} Z^i \\ \dot{x}^d \\ D_{out} \\ u \end{bmatrix}. \qquad (2.52)$$

This matrix, composed of 0, -1, $+1$, m or $1/m$, r or $1/r$ or their products, characterizes the system architecture. The S-matrix represents relations between effort and flow in the junction structure, and its form does not depend on the linearity of the model. The presence of MTF and MGY with non constant moduli introduces nonlinear coefficients m_i or r_j which may depend on the state variables as we shall see later. The S_{11} and S_{33} matrices are antisymmetric and we have:

$$S_{31} = -S_{13}^T. \qquad (2.53)$$

Thus, if we consider the case where elements (I, C) in derivative causality are not causally linked to each other (which corresponds to $S_{22} = 0$), are not linked to R-elements ($S_{23} = 0$ and $S_{32} = 0$), and are not connected to any source ($S_{24} = 0$), equation (2.52) becomes:

$$\dot{x}^i = S_{11} Z^i + S_{12} \dot{x}^d + S_{13} D_{out} + S_{14} u,$$
$$D_{in} = S_{31} Z^i + S_{33} D_{out} + S_{34} u, \qquad (2.54)$$
$$Z^d = S_{21} Z^i,$$

and if we express variables using relations (2.51), we have:

$$\dot{x}^i = S_{11} F^i(x^i) + S_{12} \dot{x}^d + S_{13} L(D_{in}) + S_{14} u,$$
$$D_{in} - S_{33} L(D_{in}) = S_{31} F^i(x^i) + S_{34} u, \qquad (2.55)$$
$$F^d(x^d) = S_{21} F^i(x^i).$$

In the most general case, these expressions do not lead to an analytic form for the state equation. In the particular case where R-elements are linear, the matrix function F^d invertible, and the expression $F^{d-1}[S_{21} \, F^i(x^i)]$ differentiable with respect to time, expression

(2.55) becomes:

$$\dot{x}^i = \left[S_{11} + S_{13}L(I - S_{33}L)^{-1}S_{31}\right]F^i(x^i) + S_{12}\frac{d}{dt}\left[F^{d-1}\left[S_{21}F^i(x^i)\right]\right]$$
$$+ \left[S_{14} + S_{13}L(I - S_{33}L)^{-1}S_{34}\right]u, \tag{2.56}$$
$$x^d = F^{d-1}\left[S_{21}F^i(x^i)\right],$$

on the condition that $(I - S_{33}L)$ is not singular. The thus obtained state model, including dependent state variables x^d, is a singular model.

In the most complex cases (elements and structure are nonlinear), it may appear to be impossible to obtain a state model in the form $\dot{x} = f(x, u)$, and the equations have to be solved numerically.

Let us reconsider the previous electrical example (fig. 2.37) and its bond-graph model (fig. 2.38) without any derivative element. The different vectors are:

$$D_{in} = \begin{bmatrix} f_3 \\ e_5 \end{bmatrix}, \quad D_{out} = \begin{bmatrix} e_3 \\ f_5 \end{bmatrix} = \begin{bmatrix} \psi_{R_1} & 0 \\ 0 & \psi_{R_2}^{-1} \end{bmatrix}(D_{in}) \quad x^i = \begin{bmatrix} p_2 \\ q_6 \end{bmatrix},$$
$$Z^i = \begin{bmatrix} f_2 \\ e_6 \end{bmatrix} = \begin{bmatrix} \psi_L & 0 \\ 0 & \psi_C \end{bmatrix}(x^i), \quad u = [e_1], \tag{2.57}$$

and the junction structure matrix, with its constant coefficients, can be written as:

$$\begin{bmatrix} e_2 \\ f_6 \\ \cdots \\ f_3 \\ e_5 \end{bmatrix} = \begin{bmatrix} 0 & -1 & \vdots & -1 & 0 & \vdots & 1 \\ 1 & 0 & \vdots & 0 & -1 & \vdots & 0 \\ \cdots & \cdots & \vdots & \cdots & \cdots & \vdots & \cdots \\ 1 & 0 & \vdots & 0 & 0 & \vdots & 0 \\ 0 & 1 & \vdots & 0 & 0 & \vdots & 0 \end{bmatrix} \begin{bmatrix} f_2 \\ e_6 \\ \cdots \\ e_3 \\ f_5 \\ \cdots \\ e_1 \end{bmatrix}. \tag{2.58}$$

This allows the definition of the matrices S_{ij} and leads to the state equation (2.46).

Consider now the bond-graph model (fig. 2.49) associated with the elastic pendulum studied previously (fig. 2.12). The elements are linear while the junction structure is nonlinear. The different vectors are:

$$x^i = \begin{pmatrix} p_3 \\ p_2 \\ q_5 \end{pmatrix} = \begin{pmatrix} p_x \\ p_y \\ r \end{pmatrix}, \quad Z^i = \begin{pmatrix} f_3 \\ f_2 \\ e_5 \end{pmatrix} = \begin{bmatrix} 1/m & 0 & 0 \\ 0 & 1/m & 0 \\ 0 & 0 & k_1 \end{bmatrix} \begin{pmatrix} p_x \\ p_y \\ r - L_0 \end{pmatrix}, \tag{2.59}$$

where L_0 is the length of the unloaded spring:

$$D_{in} = [f_4], \quad D_{out} = [e_4] = [R]D_{in}, \quad u = [e_1]. \tag{2.60}$$

The junction structure matrix can be written:

$$\begin{pmatrix} e_3 \\ e_2 \\ f_5 \\ \cdots \\ f_4 \end{pmatrix} = \begin{bmatrix} 0 & 0 & -\sin\theta & \vdots & -\dfrac{\cos\theta}{r} & \vdots & 0 \\ 0 & 0 & -\cos\theta & \vdots & \dfrac{\sin\theta}{r} & \vdots & 1 \\ \sin\theta & \cos\theta & 0 & \vdots & 0 & \vdots & 0 \\ \cdots & \cdots & \cdots & \vdots & \cdots & \vdots & \cdots \\ \dfrac{\cos\theta}{r} & -\dfrac{\sin\theta}{r} & 0 & \vdots & 0 & \vdots & 0 \end{bmatrix} \begin{pmatrix} f_3 \\ f_2 \\ e_5 \\ \cdots \\ e_4 \\ \cdots \\ e_1 \end{pmatrix}. \tag{2.61}$$

The S-matrix includes nonlinear terms which depend not only on the state variable r, but also on the angle θ. Even though θ is not a state variable, it should be included in the state vector since its value must be known at each calculus step in the simulation phase. The complete differential model, associated with the bond-graph model (fig. 2.49) of the elastic pendulum, and deduced from expressions (2.51) and (2.52), with $S_{12} = 0$ and $S_{2i} = 0 \ (i = 1, \ldots, 4)$, is then:

$$\dot{p}_3 = \dot{p}_x = -R\frac{\cos^2\theta}{r^2}\frac{p_3}{m} + R\frac{\sin\theta\cos\theta}{r^2}\frac{p_2}{m} - \sin\theta k_1(r - L_0),$$

$$\dot{p}_2 = \dot{p}_y = \frac{R}{r^2}\sin\theta\cos\theta\frac{p_3}{m} - \frac{R}{r^2}\sin^2\theta\frac{p_2}{m} - \cos\theta k_1(r - L_0) + mg, \qquad (2.62)$$

$$\dot{q}_5 = \dot{r} = \sin\theta\frac{p_3}{m} + \cos\theta\frac{p_2}{m},$$

with:

$$\dot{\theta} = \frac{\cos\theta}{r}\frac{p_3}{m} - \frac{\sin\theta}{r}\frac{p_2}{m}. \qquad (2.63)$$

The model is then in the following form:

$$\dot{x} = A(x, \theta)x + Bu,$$
$$\dot{\theta} = f(x, \theta). \qquad (2.64)$$

It is always possible to show up such a matrix $A(x, .)$ when only the junction structure is nonlinear.

2.2.5 Structural properties for a system deduced from its bond-graph model

A property is called "**structural**" if it depends only on the type of elements which compose the system and on the way they are interconnected, and if it remains true for almost all the particular numerical values of the parameters.

The properties presented here are demonstrated in Sueur and Dauphin-Tanguy (1991). Some other structural properties (such as identifiability and discernibility) are given in chapter 3.

2.2.5.1 Invertibility of the state matrix associated with a bond-graph

The state matrix concept does not have a meaning except in the linear domain. Thus, a linear state model, under the form $\dot{x} = Ax + Bu$, is characterized by its minimal order which is the A-matrix dimension, as well as by the rank of A which corresponds to the number of eigenvalues different from zero. However, when the model is written as $\dot{x} = A(x, .)x + B(.)u$, which is the case for state models issued from bond-graphs with linear elements and nonlinear junction structure (refer to the elastic pendulum example), it may appear interesting to determine if the rank of $A(x)$ is structurally maximal.

2.2.5.1.1   Linear elements and structure

Definition 1 *The minimal order of a model, that is the state matrix dimension, is equal to the number $(n_i)_i$ of I- and C-elements in integral causality when an integral causality is assigned to the bond-graph.*

Theorem 1 *The **structural rank** q **of the state matrix** associated with a bond-graph model initially in integral causality is equal to the number of I- and C-elements in integral causality accepting a derivative causality when a derivative causality is assigned to the bond-graph. If $(n_i)_i$ (respectively $(n_i)_d$) represents the number of I- and C-elements in integral causality when an integral causality i (respectively derivative d) is applied on the bond-graph, we have:*

$$q = (n_i)_i - (n_i)_d.$$

As an example, consider the following bond-graph:

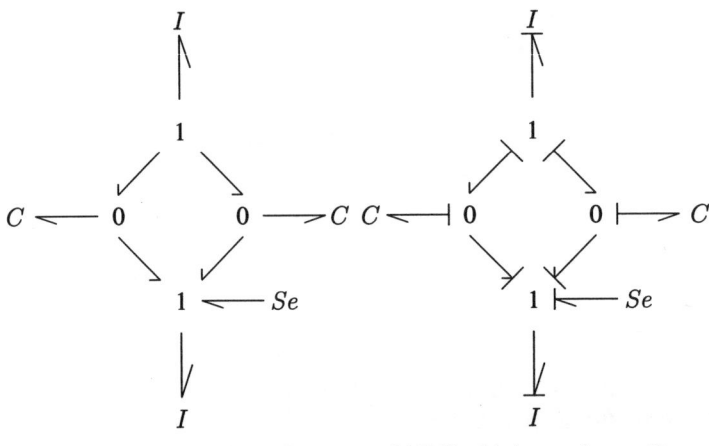

(a) Acausal bond-graph   (b) BG with integral causality

Figure 2.60: Bond-graph associated with a fourth-order system.

The four I- and C-elements admit an integral causality without creating any conflict (fig. 2.60 (b)), the model order is $n = (n_i)_i = 4$.

If a derivative causality is assigned to all I- and C-elements, a non solvable causal loop (a causality loop which does not allow explicit derivation of the relations) (fig. 2.61 (a)) appears. This is not allowable and the choice to assign a complete derivative causality is not possible. We show that the unique way to avoid conflicts is to keep one I and one C in integral causality (fig. 2.61 (b)), that is $(n_i)_d = 2$.

The structural rank of the state matrix is then $q = 2$.

2.2.5.1.2   Linear elements and nonlinear structure

The procedure announced above for the linear case leads, by inverting the causalities, to determine whether equation $\dot{x} = Ax + Bu$, deduced from an integral causality assignment

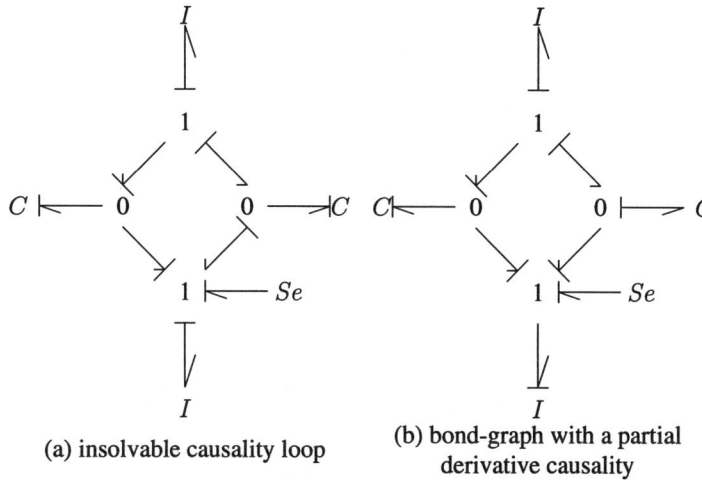

(a) insolvable causality loop

(b) bond-graph with a partial derivative causality

Figure 2.61: System of rank 2.

for I and C, can be transformed to the form:

$$x = A^{-1}\dot{x} - A^{-1}Bu,$$

corresponding to the result of a derivative causality assignment. When all I- and C-elements may admit a derivative causality without any conflict, it is equivalent to say that A is invertible and its rank is maximum.

When the state model is under the form $\dot{x} = A(x,.)x + B(.)u$, it is interesting to use the procedure presented in the following for the elastic pendulum example.

The bond-graph model of the pendulum, in integral causality in figure 2.49, leads to a state vector x with three components. When the derivative causality is assigned to the I- and C-elements, as shown figure 2.62, no conflict appears.

$A^{-1}(x, \theta)$ exists **structurally**.

In order to check the result, let us calculate the determinant of $A(x, \theta)$, with $A(x, \theta)$ deduced from equation (2.63) in the form:

$$A(r, \theta) = \begin{bmatrix} -\dfrac{R\cos^2\theta}{m} \dfrac{1}{r^2} & \dfrac{R\sin\theta\cos\theta}{m} \dfrac{1}{r^2} & -k_1\sin\theta \\ \dfrac{R\sin\theta\cos\theta}{m} \dfrac{1}{r^2} & -\dfrac{R\sin^2\theta}{m} \dfrac{1}{r^2} & -k_1\cos\theta \\ \dfrac{\sin\theta}{m} & \dfrac{\cos\theta}{m} & 0 \end{bmatrix}, \qquad (2.65)$$

then:

$$\Delta(r) = \det A(r, \theta) = -\dfrac{k_1 R}{m^2 r^2}. \qquad (2.66)$$

In this particular case, it appears that $\Delta(r)$ is independent of θ, but not constant since it depends on the state variable r. $\Delta(r)$ is always different from zero and bounded since r, associated with the length of the pendulum bar, is in the interval $[L_o, r_{\max}]$.

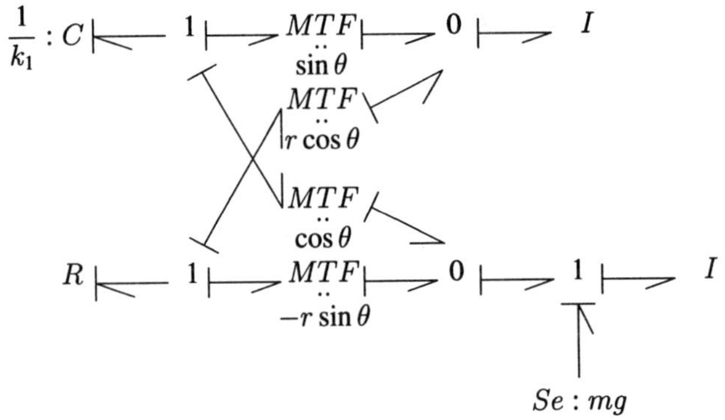

Figure 2.62: Bond-graph model of an elastic pendulum in derivative causality.

The structural rank of $A(r, \theta)$ is then always maximal and equal to 3. Also, in this case, it corresponds exactly to the real rank since this property is true whatever the values of r and θ.

Remark 6

The extension of this procedure to more complex cases, with nonlinear elements for instance, needs to be carried out carefully.

2.2.5.2   Structural controllability and observability

A very general theorem was proposed by Sueur and Dauphin-Tanguy (1991) in the linear case. It uses the concept of dualization for sources and detectors, which consists in transforming a source (or detector) of effort (respectively of flow) into a source (or detector) of flow (respectively of effort). Demonstrated by using the junction structure of the system, this theorem can be interpreted graphically as follows:

Theorem 2.1 *A linear system modeled by a bond-graph is structurally controlable (respectively observable) if and only if:*

1. *A causal path exists linking a source (respectively a detector) to each dynamic element I and C in integral causality when an integral causality is assigned to the bond-graph.*

2. *All I- and C-elements admit a derivative causality when a derivative causality is assigned to the bond-graph, by using eventually a dualization of the sources (respectively the detectors).*

The extension to the nonlinear case of the procedure summed up by theorem 2.1 is presently under study.

2.2.6   Bond-graph study of the existence of a law for input-output decoupling-linearization

When the state model representing the physical system is written in the form:

$$\dot{x} = f(x) + \sum_{i=1}^{m} u_i g_i(x),$$
$$y = h(x), \qquad\qquad\qquad (2.67)$$

an algorithm, based on the construction of the associated system-graph, enables a solution to the input-output decoupling problem. The main definitions and results of the method are developed in volume III of this series.

This graphic approach is also possible using the graph of the causal paths (Maschke, 1990), defined from the bond-graph for systems with **nonlinear structure** and **linear elements** without any dissipation phenomena.

2.2.6.1   Graph of the causal paths

2.2.6.1.1   Definition

The graph of the causal paths, issued from a bond-graph model, consists of:

- nodes: I- and C-elements (state nodes), sources (input nodes) and detectors (output nodes);
- arcs: any causal path defined previously (simple, mixed direct or indirect) or generalized (concatenation of a signal bond with a causal path for an MTF-element the modulus of which is equal to $\cos\theta$ for example, where θ is re-evaluated at each step of the calculus).

The junction structure, which includes all information about the causal paths, is used to construct the graph of the causal paths. To be able to include the input-output relations, definition (2.52) of the S-matrix must be extended so as to take into account the output, so:

$$
\begin{bmatrix} \dot{x}^i \\ Z^d \\ D_{\text{in}} \\ y \end{bmatrix}
=
\begin{bmatrix}
S_{11} & S_{12} & S_{13} & S_{14} \\
S_{21} & S_{22} & S_{23} & S_{24} \\
S_{31} & S_{32} & S_{33} & S_{34} \\
S_{41} & S_{42} & S_{43} & S_{44}
\end{bmatrix}
\begin{bmatrix} Z^i \\ \dot{x}^d \\ D_{\text{out}} \\ u \end{bmatrix}.
$$

The class of considered systems does not include dissipation elements, all the S_{3i} and S_{i3}, $i = 1, \ldots, 4$, are equal to zero.

Also, the previous hypotheses are maintained: I- and C-elements in derivative causality are connected neither to each other ($S_{22} = 0$) nor to the sources ($S_{24} = 0$).

2.2.6.1.2   Graph construction

Each arc is associated with the coefficients, or their derivatives, of the sub-matrices S_{ij} of the junction structure S.

1. An arc (called a differential arc) between an input node u_j, associated with a source, and a state node x_k^i, exists if the corresponding coefficient $S_{14}(k, j)$ is different from zero. This corresponds to the existence of a causal path between the source $Se : u_j$ and the dynamic element I or C in integral causality associated with x_k^i.

2. A (differential) arc between the state node x_l^i and the state node x_k^i exists if the terms $S_{11}(k, l)$ or $d/dt[S_{21}(k, m)]$ for at least m element of $\{1, \ldots, n_i\}$, are different from zero (n_i= number of I and C in integral causality). This corresponds to a causal connexion between the two associated dynamic elements I or C.

3. An (algebraic) arc between the state node x_l^i and the state node x_k^d exists if the coefficients $S_{12}(l, k)$ and $S_{21}(k, l)$ are different from zero. This corresponds to a causal connexion between a dynamic element in integral causality and another in derivative causality.

4. An (analytic) arc between a state node x_k^i and the output node y_l exists if the term $S_{41}(l, k)$ is different from zero. This corresponds to the existence of a causal path between the dynamic element I or C in integral causality and the detector indexed l.

2.2.6.1.3 *Comparison between the graph of the causal paths and the system-graph*

The graph of the causal paths is purely graphically deduced from the bond-graph. The system-graph is deduced from the Jacobian of the analytic function, which describes the dynamics of the independent energy variables in the model (2.67).

However, these two graphs are comparable from the point of view of their interpretation in terms of dynamical system equations. The graph of the causal paths is associated with the differential and algebraic equation system:

$$\begin{cases} \dot{x}^i = S_{11}F^i x^i + S_{12}\dot{x}^d + S_{14}u, \\ x^d = F^{d-1}S_{21}F^i x^i, \\ y = h(x^i). \end{cases} \tag{2.68}$$

The system-graph is, by definition, associated with the analytic system linear for the input, from which ordinary differential equations can be deduced from the previous equation and are written under the form:

$$\dot{x}^i = \left[I_d - S_{12}F^{d-1}\left(S_{21} + \frac{\partial S_{21}}{\partial x^i} \right) F^i \right]^{-1} (S_{11}F^i x^i + S_{14}u), \tag{2.69}$$
$$y = h(x^i).$$

A graphic transformation of the graph of the causal paths into a system-graph represents thus equivalently the graphic elimination of the dependent energy variables x^d in equation (2.68), leading to (2.69).

2.2.6.2 Structural graphic diagnosis in the graph of the causal paths

2.2.6.2.1 Relative indices

Definition 2

- *The length of an arc, noted l^c, is equal to 1 for a differential arc, and 0 for an algebraic or analytic arc;*

- *The length of a path is equal to the sum of the lengths of the arcs which compose it;*

- *The deviation between two nodes, noted $e(.,.)$ is the minimal length for an elementary path (without a cycle) oriented from the starting to the ending node.*

Proposition 1 *The relative indices admit as a minimal value the deviation between the whole set of input nodes and the output node y_i in the graph of the causal paths:*

$$\forall i \in (1, \ldots, m), \quad r_i \geq e^c(u, y_i) = e_i^c. \tag{2.70}$$

Remark 7

The minimal values deduced from the system-graph and the graph of the causal paths are similar.

2.2.6.2.2 Rank of the decoupling matrix

Definition 3 (minimal graph)
*The **minimal graph** extracted from the graph of the causal paths, is the partial graph formed by all arcs composing the minimal paths from the whole set of the inputs to the outputs.*

Proposition 2
● *A structural solution exists for the inputs-outputs decoupling-linearization problem for systems whose graph of the causal paths verifies $r_i = e_i^c$, $i = 1, \ldots, m$, if and only if the minimal graph admits a complete elementary coupling between inputs and outputs.*

● *The rank of the decoupling matrix (volume III of this series) is inferior to the maximal size of an elementary input-output coupling in the minimal graph issued from the graph of the causal paths.*

2.2.7 Numerical problems

Some numerical problems may occur when a bond-graph model includes dynamic elements in derivative causality or algebraic loops. The determination of particular causal paths allows an *a priori* detection of these difficulties. Indeed:

- a causal loop between two R-elements (called an "algebraic loop") leads to implicit equations and then to risks of nonconvergence of the simulation algorithm;

- a causal loop between two I-elements (or two C-elements) not going through a GY, points out to the fact that one of the I- (or C-) ports is in derivative causality, and then to the existence of a state model under a differential and algebraic equation system.

One way to eliminate these difficulties is, as mentioned before, to introduce new elements (R, C, or I), which "break" these causal paths by the introduction of new causalities. As far as possible, a physical justification for these new components must be discovered (friction neglected initially, elasticity in an articulation, compressibility of a fluid) and the parameters' value must be chosen in such a way that the disturbances on the system behavior are very weak.

2.3   Applications

2.3.1   Modeling of converters in power electronics

The originality of power electronics systems is due to the presence of nonlinear components with two possible states ("on" or "off"), which lead to different states of conduction for the system and then to multiple models between which commutations occur. Another particularity of these systems is that they are composed of different devices as source, load, converter, controller.

The modeling approaches chosen for static converters are closely dependent on the desired purposes. From an "electrical engineering" point of view, two kinds of purposes can be defined, which lead to different constraints and possibilities of simplification:

- when the study concerns the converter alone, then no combination of states of the components can be excluded, but its environment can be simplified;
- when the study concerns the whole converter-machine assembly, the converters operation may, in general, be considered as known, but the complete system must be taken into consideration.

2.3.1.1   "Classical" methods of modeling

2.3.1.1.1   "Constant topology" method

These methods are based on the representation of the semi-conductors by impedances (usually resistors, sometimes connected to inductances) the value of which is binary, small (sometimes null) when the component is on, large when the component is off. The converter is thus represented then as a gridded network including the source and the load, and the classical methods for circuit analysis remain valid for state model construction, under a unique form. The different conduction states are taken into account just by modifying the elements of the state matrices and of the control.

The inconveniences of this type of modeling are the following:

- All the power system elements (source - converter - load) must be represented by circuit elements.

- Numerical problems appear for the resolution of the equations. Indeed, the classical methods for numerical resolution of differential equations (type Runge-Kutta) are "step-by-step" methods, with a time interval chosen depending on the time constant of the system studied. Choosing the impedance with a large value when the component is "off" leads to "stiff" equations, with the association of very fast and very slow phenomena which are difficult to take into account simultaneously.

  This type of methods is better adapted for the study of isolated converters.

2.3.1.1.2 *"Variable topology" methods*

These methods are based on the representation of semi-conductors by perfect switches. The whole set of source - converter - load appears under a gridded network form, with a number of meshes and nodes depending on the state of the different components.

This type of modeling has the advantage that it suppresses the parasitical time constants of the previous methods, but presents some inconveniences:

- the problem of modeling and representing the machines by circuit elements remains;
- when the converter under consideration includes n components which may commute independently, there are 2^n possible topologies and then 2^n models.

2.3.1.1.3 *"Restricted topology" method*

This method is based, as previously, on the representation of the switching devices by ideal switches, differs, however, by the choice of a set of possible topologies. The whole set of source-converter is then modeled by equivalent diagrams which show explicitly the supply voltage of the machines and permits the use, directly or indirectly (after the Park transformation for example) of their steady state equations.

The difficulties involved with this methods are:

- the necessary knowledge *a priori* of the possible topologies;
- they are not adapted to the study of the converter itself or at least to the search of its working mode.

2.3.1.1.4 *Mean Value Method*

This method, used mainly to find out control laws, consists in representing the behavior of the converter using the mean value of its output. The problems involved with this method are:

- the resulting error of such an approximation is difficult to determine;
- this approximation is valuable only when the power of the controled machines is weak.

2.3.1.2   Bond-graph modeling

2.3.1.2.1   *Modeling of the elementary converters*

The elementary converters (diode, thyristor, commutations transistor, etc.) are all characterized by a nonlinear current-voltage law. The difference intervenes at the control level for the two states on or off. In fact, a diode is not controlable since the change of state depends uniquely on its environment. The thyristor is controlable for its switching on thanks to the trigger current but not in the switching off. A commutation transistor is controlable both ways.

The characteristic curves of these components are represented in figure 2.63 (a), (b) and (c). We introduce the Boolean parameter m to represent the state ON ($m = 1$) and the state OFF ($m = 0$), with the commutation logic for these two switches given figure 2.64 (a), (b) and (c). The slope $1/R$ of the linear part is generally considered as infinite since it is usual, in order to simplify the study of components and their association, to assume the devices are ideal.

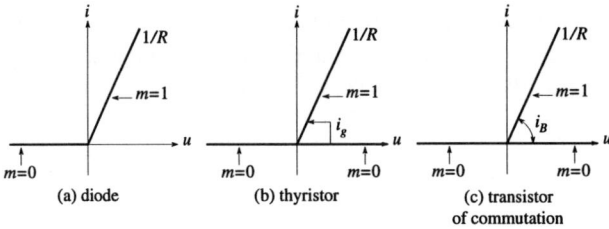

Figure 2.63: Characteristic curves of three elementary power electronic converters.

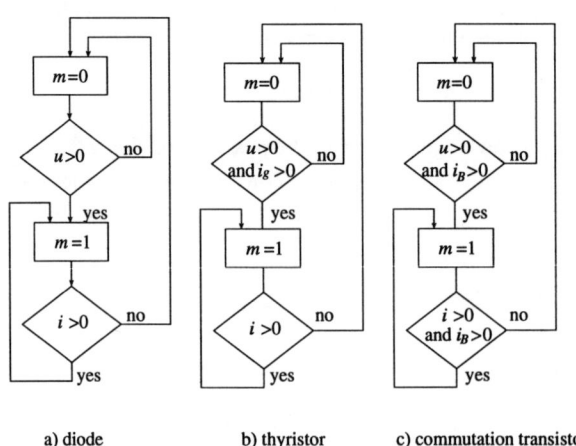

Figure 2.64: Commutation logic for three elementary converters.

A type changing of the component modifies the logic associated with the Boolean m. The evolution law for this parameter can be refined to take into account non instantaneous commutation phenomena and allows then a more detailed study of the commutation transient states.

These remarks led us to define the bond-graph model of an elementary switch as the association of an element $R : R_c$ and a modulated transformer MTF, the Boolean modulus of which is noted m.

The resistance value R_c of the switch may be very small if the user thinks the component is ideal. The characteristic curve of the switching ON of the model can be chosen nonlinear, for instance parabolic, for a more precise modeling.

2.3.1.2.2 *Modeling of an inverter*

Consider the system represented in figure 2.65.

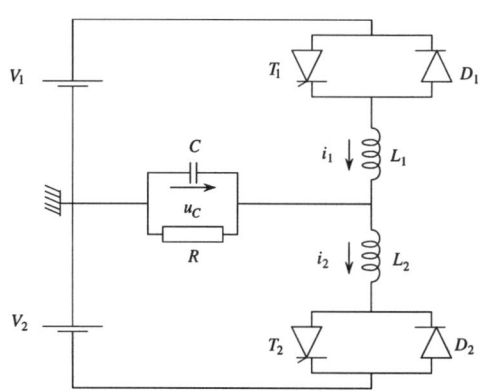

Figure 2.65: Diagram of a voltage converter.

This example is studied in the paper written by Le Dœuff (1989), using two classical methods, the first with a constant topology, the other with a variable topology and supposing that the switches T_i and D_i at the same branch of the circuit cannot conduct simultaneously. This hypothesis is confirmed by the study of the bond-graph presented in figure 2.66. When causalities are assigned with the hypothesis that the switches are ideal, and that the causal path gains between the sources and the dynamical elements I and C are independent of the switches, a causality conflict arises at the two 0-junctions represented by a dotted square which shows, in fact, that the two branches (via the diode and the thyristor) cannot be active simultaneously (fig. 2.66).

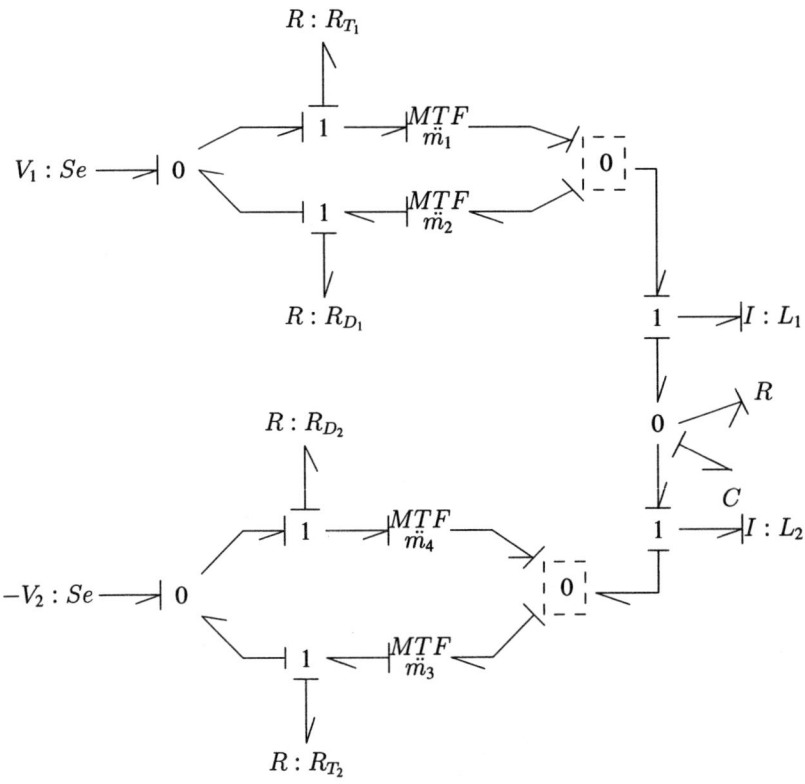

Figure 2.66: Bond-graph of the convecter of figure 2.65.

The state equation associated with the bond-graph model is obtained as explained above in the following form:

$$
\begin{bmatrix} \dot{p}_{L_1} \\ \dot{p}_{L_2} \\ \dot{q}_C \end{bmatrix} = \begin{bmatrix} -m_1^2\dfrac{R_{T_1}}{L_1} - m_2^2\dfrac{R_{D_1}}{L_1} & 0 & -\dfrac{1}{C} \\ 0 & -m_3^2\dfrac{R_{T_2}}{L_2} - m_4^2\dfrac{R_{D_2}}{L_2} & \dfrac{1}{C} \\ \dfrac{1}{L_1} & -\dfrac{1}{L_2} & -\dfrac{1}{RC} \end{bmatrix} \begin{bmatrix} p_{L_1} \\ p_{L_2} \\ q_C \end{bmatrix}
$$
$$
+ \begin{bmatrix} m_1 + m_2 & 0 \\ 0 & m_3 + m_4 \\ 0 & 0 \end{bmatrix} \begin{bmatrix} V_1 \\ -V_2 \end{bmatrix},
$$

$$
y = u_C = \begin{bmatrix} 0 & 0 & \dfrac{1}{C} \end{bmatrix} \begin{bmatrix} p_{L_1} \\ p_{L_2} \\ q_C \end{bmatrix},
$$

where p_{L_1}, p_{L_2} are the flux linkage in the inductances and q_C is the charge in the capacitor.

The state model is nonlinear, the control variables u_1 and u_2 correspond to the Booleans m_1 and m_3 (or m_1^2 and m_3^2, which is equivalent) associated with thyristors. Equation (2.71) can thus be written as:

$$\dot{x} = (A_0 x + B_0 u_0) + (A_1 x + B_1 u_0) u_1 + (A_2 x + B_2 u_0) u_2,$$

with:

$$u_0 = \begin{bmatrix} V_1 \\ -V_2 \end{bmatrix}, \quad u = \begin{bmatrix} u_0 \\ u_1 \\ u_2 \end{bmatrix}, \quad B_0 = \begin{bmatrix} m_2 & 0 \\ 0 & m_4 \\ 0 & 0 \end{bmatrix}, \quad B_1 = \begin{bmatrix} 1 & 0 \\ 0 & 0 \\ 0 & 0 \end{bmatrix},$$

$$B_2 = \begin{bmatrix} 0 & 0 \\ 0 & 1 \\ 0 & 0 \end{bmatrix},$$

$$A_0 = \begin{bmatrix} -m_2^2 \dfrac{R_{D_1}}{L_1} & 0 & -\dfrac{1}{C} \\ 0 & -m_4^2 \dfrac{R_{D_2}}{L_2} & \dfrac{1}{C} \\ \dfrac{1}{L_1} & -\dfrac{1}{L_2} & -\dfrac{1}{RC} \end{bmatrix}, \quad A_1 = \begin{bmatrix} -\dfrac{R_{T_1}}{L_1} & 0 & 0 \\ 0 & 0 & 0 \\ 0 & 0 & 0 \end{bmatrix}, \qquad (2.72)$$

$$A_2 = \begin{bmatrix} 0 & 0 & 0 \\ 0 & -\dfrac{R_{T_2}}{L_2} & 0 \\ 0 & 0 & 0 \end{bmatrix}.$$

It is interesting to notice that the **obtained model is unique**, and includes all possible topologies for the system. The Booleans m_1 and m_2 (respectively m_3 and m_4) can be simultaneously null which disactivates the branch indexed 1 (respectively 2) of the circuit, but cannot be simultaneously equal to 1 since the two switches cannot conduct at the same time, as has been shown by the causality conflict.

The terms in the diagonal of the matrices A_0, A_1 and A_2 include the resistances of the switches which can be considered null ("classical" approach), or can be chosen much weaker than the circuit's resistances. The deviation when neglecting the resistances may be important, as shown in the following table which gives the eigenvalues of the state matrix $A(m_i)$ (equation (2.71)) for certain working points when $L_1 = L_2 = 76\mu H$, $R = 50\Omega$, and $C = 3.33\mu F$.

The results obtained for null resistances are identical to those given in Le Dœuff (1989).

We note that when a single branch of the circuit is active, the eigenvalues are slightly influenced by the resistances of the switches, which is not the case when the two branches are active. A very fast mode appears for values of the switches' resistances which are much weaker than the load's resistance.

2.3.1.3 Conclusion

Systematic methods for graph modeling of converters were proposed by Buyse (1984) for flow-graphs, and by Manesse (1987) for Petri-nets. They necessitate that the analyst explores different possible configurations.

The interest of the bond-graph approach for such a kind of problems are:

Case	Resistances of the switches	Eigenvalues of A
$(m_1$ or $m_2) = 1$	$0\,\Omega$	0 and $-3 * 10^3 \pm 6.28 * 10^4 i$
$m_3 = m_4 = 0$	$0.01\,\Omega$	0 et $-3.07 * 10^3 \pm 6.28 * 10^4 i$
or $m_1 = m_2 = 0$	$0.1\,\Omega$	0 and $-3.66 * 10^3 \pm 6.27 * 10^4 i$
$(m_3$ or $m_4) = 1$	$1\,\Omega$	0 and $-9.58 * 10^3 \pm 6.27 * 10^4 i$
	$0\,\Omega$	0 and $-3 * 10^3 \pm 8.88 * 10^4 i$
$(m_1$ or $m_2) = 1$	$0.01\,\Omega$	$-1.316 * 10^2$ $-3.07 * 10^3 \pm 8.88 * 10^4 i$
et $(m_3$ or $m_4) = 1$	$0.1\,\Omega$	$-1.32 * 10^3$ $-3.66 * 10^3 \pm 8.89 * 10^4 i$
	$1\,\Omega$	$-1.32 * 10^4$ $-9.58 * 10^3 \pm 8.89 * 10^4 i$

- bond-graph includes the analysis hypothesis (switches supposed perfect or not);
- the method can be applied for complex converters and allows a direct definition of the possible topologies under a unique model form;
- bond-graph models can be coupled together with bond-graph models of machines and loads which leads to a complete model for the system (Ducreux et al., 1992), which is an improvement in comparison with the "classical" approaches.

2.4   Pneumatic part of the electropneumatic driving systems

2.4.1   System under consideration

The system under consideration consists of a test prototype made in order to verify the feasibility of different control laws either of position or of effort by experimental study. Before developing this test model, a mathematical model was established to validate the proposed control laws and to facilitate the understanding of the dynamic behavior. The system under consideration (fig. 2.67) is composed of a single rod, double acting, linear pneumatic actuator. This actuator drives a load carriage which runs on ball-bearings and hardened steel rods. The load mass can be increased by adding one or several additional masses. A single-stage four-way electropneumatic servovalve controls the power which flows from the supply to the actuator. By using information given by position, velocity, acceleration and pressure sensors, the position of the load carriage or the effort delivered to a fixed stop can be controled.

By examining the schema of the spool-sleeve assembly in figure 2.67 it can be seen that the servovalve flow stage is composed of four orifices whose areas are designated by A_{sp}, A_{pe}, A_{sn}, A_{ne}.

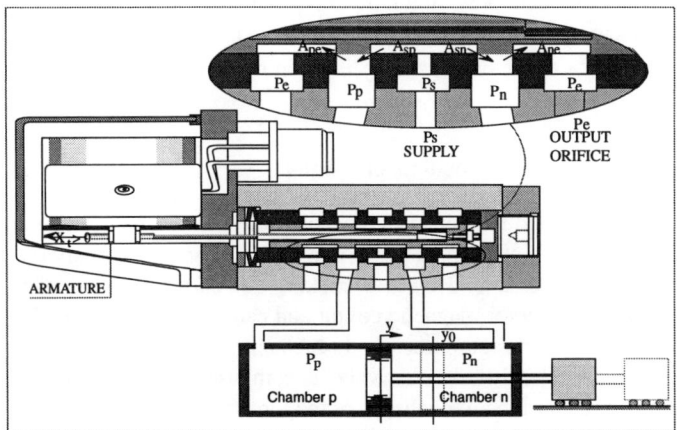

Figure 2.67: Technological schema.

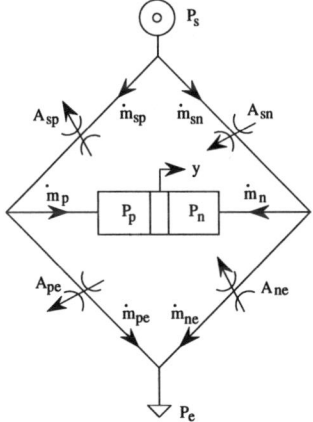

Figure 2.68: Principle scheme.

These orifices are disposed in a Wheatstone bridge lay out. As each cylinder chamber communicates with an output orifice the actuator is placed in the diagonal of this Wheatstone bridge. Figure 2.68 schematizes these preliminary technological considerations. In the rest of the section, an assembly of two orifices placed on the same side will be called a half bridge. For example, the orifices whose sections are A_{sp} and A_{pe} make up the half bridge associated to the chamber p.

2.4.2 General considerations and hypothesis

Establishing the mathematical model of the system under consideration implies taking into account:

- the spool positionning dynamic;
- the flow static characteristics i.e. the flow laws of the four orifices, including in the proximity of the spool to the closed postion. For this configuration, the aim is to obtain the reproduction by simulation of the changes in the chamber pressures P_p and P_n, the output orifices being assumed closed;
- the specificities of the pneumatic cylinder in terms of thermal exchanges and frictions.

2.4.2.1   The dynamic of the spool positionning

The most rigorous method for taking into account the servovalve positionning dynamic consists of establishing a state model as proposed by Lebrun and Scarvarda (1979) for an electrohydraulic servovalve. However, making this choice realistic leads to detailed modeling of the torque motor magnetic circuit and can only be justified when the aim is servovalve modeling for design purposes. In the context of the example presented here, it is preferable to consider that the servovalve is composed of two subsystems. The first, represented by a transfer function consists of the torque motor-spool assembly. The four orifices of the servovalve flow stage constitute the second subsystem which will be modeled by bond-graphs.

Logically the transfer function characterizing the spool positioning dynamic must relate the spool position to the servovalve input current. This choice implies the introduction of a flow law $\dot{m} = f(i, P)$ for each of the four orifices. To overcome this difficulty (Det et al., 1988), have proposed to take into account the spool positioning dynamic by introducing this dynamic between the servovalve input current i and a virtual current i_v. This solution enables the introduction of experimental flow laws in the model. The transfer function is usually deduced from the frequency response obtained by choosing the servovalve current as input and the spool position as output.

2.4.2.2   Pneumatic part

In the prototype, the distance between a variable orifice and the piston does not exceed 300 mm; this corresponds to a time of one millisecond for pressure wave travel. Therefore we can neglect the line effects and suppose that the variables are homogeneous in each of the two chambers. Moreover the power necessary to move each of the air columns is negligible with regard to the circuit dimensions and the maximal value of the acceleration. Thus the modeling of the pneumatic part is reduced to an orifice and chamber assembly.

This modeling must take into account, on the one hand the fluid compressibility, and on the other hand the fact that a chamber is an open system, i.e. a system containing a variable mass of gas. The analysis of an open system is usually made by adopting Euler's point of view, which consists of considering a fixed reference volume and of establishing for this volume, a mass balance, a momentum balance and an energy balance. The translation of this point of view, into bond-graphs implies putting one self in the generalized bond-graphs context, as shown by Breeveld (1984). This type of approach derives from the classical bond-graphs and we therefore adopt Karnopp's (1979) point of view for modeling the pneumatic part. This point of view consists of making a pseudo bond-graph by introducing pseudo bonds for which the product of effort and flux variables associated to a bond does not have the dimension of a power. These pairs of effort and flux are chosen in order to facilitate the translation of the different balances. Karnopp uses the following pairs: temperature-energy flow rate $(T - \dot{E})$, pressure-mass flow rate $(P - \dot{m})$.

It can be noticed, besides the fact that the dimension of the effort and flow variables product is not correct, that the classical properties of a true bond-graph (sign convention, causality assignment, classification into R and C elements) are conserved in a pseudo bond-graph. In this context the word bond-graph of the pressure source-flow stage-cylinder-load carriage assembly takes the following form:

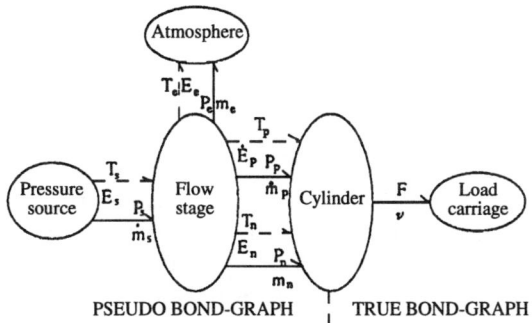

Figure 2.69: Word bond-graph.

2.4.3 Modeling the pneumatic part

As well as the choice of the pseudo bonds $(T-\dot{E})$ and $(P-\dot{m})$, Karnopp (1979) introduced a four-ports R-pseudo element and three-ports C-pseudo element modeling respectively a variable area orifice and a chamber. We present the two pseudo elements before modeling the pneumatic part.

2.4.3.1 Cylinder chamber

One associates (Breeveld, 1984) one multiport C-element to a energy storing element for which the storage energy expression is a function of several displacement variables. To each displacement variable corresponds a port whose flux variable is the time derivative of this displacement variable and whose effort variable is the partial derivative of the energy with respect to the displacement variable under consideration.

For example, for a capacitor with a moving plate the stored energy $E = q^2/2C(x)$ is a function of two displacement variables, q (charge) and x (plate displacement). The bond-graph of this capacitor is represented by a two-port C-element. The electrical port corresponds to the displacement variable q, the mechanical port is related to the mechanical displacement following the bond-graph below (fig. 2.70):

If the kinetic energy of the gas contained in the chamber is neglected and if it is assumed that all the variables are homogeneous, it can be shown that the gas energy is a function of the three following displacement variables: the entropy S, the mass m, and the volume V. A pneumatic chamber must be represented by a three-port C-element.

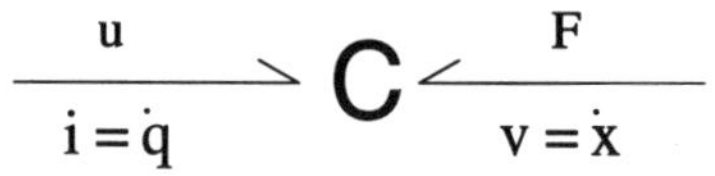

$$\xrightarrow[\;i=\dot{q}\;]{\quad u \quad} \mathbf{C} \xleftarrow[\;v=\dot{x}\;]{\quad F \quad}$$

Figure 2.70: Two port C-element representing a mobile plate capacitor.

To facilitate the representation of the different balances, Karnopp chose as displacement variables the internal energy $U = E$, the mass m and the volume V of the gas contained in the chamber, which leads to a three-port C-element. To these displacement variables Karnopp associated the following flow variables: the energy flow rate $\dot{E}$, the mass flow rate $\dot{m}$, and the volume $\dot{V}$ variation rate. Two pseudo bonds correspond respectively to the variable pairs $(T - \dot{E})$ and $(P - \dot{m})$ and constitute two of the three of the C-element ports. The third port, associated with the variable pair $(P, \dot{V})$ is a true port which will enable the connexion of the pneumatic pseudo bond-graph to the mechanical true bond-graph. In general each of the three constitutive relations of a three-port C-element relates the effort variable of the port under consideration to the three displacement variables (generalization of the constitutive relation of a simple one-port C-element). In this case two ports have the same variable effort, the pressure P. The number of constitutive relations is reduced to two. By assuming that the gas is perfect, it can be easily shown that these two relations have the following expression in integral causality:

$$P = \frac{r}{C_v} \cdot \frac{U}{V}, \qquad T = \frac{1}{C_v} \cdot \frac{U}{m}, \tag{2.73}$$

with the initial conditions:

$$m_0 = \frac{P_0 V_0}{r T_0}, \qquad U_0 = \frac{P_0 V_0 C_v}{r}, \tag{2.74}$$

with r a perfect gas constant and C_v the specific heat at constant volume.

The three-port pseudo C-element does not respect the properties of a multiport C-element from an energy point of view, the representation of the mass balance and of the energy balance in the chamber implies adding the two supplementary 0-pseudo junctions.

The mass balance in the chamber shown in figure 2.71 has the following expression:

$$\frac{dm}{dt} = \dot{m}_1 - \dot{m}_2. \tag{2.75}$$

The pseudo 0-junction associated with the bonds in continuous lines ensures the satisfaction of this mass balance. If the thermal exchange between the piston and air is neglected the energy balance is written:

$$\frac{dE}{dt} = \frac{dU}{dt} = \dot{m}_1 h_1 - \dot{m}_2 h_2 - P\dot{V} + \dot{Q}. \tag{2.76}$$

In this expression $\dot{Q}$ represents the thermal flux exchanged by convection, h_1 and h_2 respectively indicate the specific enthalpy of the air entering into the chamber and the air exhausting from the chamber.

The exhausting thermal flow $\dot{Q}$ exchanged by convection is proportional to the temperature difference between the cylinder $(Tcyl)$ and the air (T), to the lateral area A of the cylinder and to the expression:

$$\dot{Q} = kA(Tcyl - T). \tag{2.77}$$

The pseudo bond-graph representation of the thermal phenomena is made by choosing the absolute temperature as effort variable and the thermal flow $\dot{Q}$ as flow variable. The expression (2.75) corresponds to the constitutive relation of a R pseudo element. The thermal flux $\dot{Q}$ results from the connexion at constant flow $\dot{Q}$ (i.e. by mean of a pseudo 1-junction), of this R pseudo element, of a pseudo effort source associated to the cylinder and to the pseudo 0-junction which is necessary for representing the energy balance.

Besides this pseudo 0-junction, satisfying the energy balance involves the introduction of a modulated pseudo flow source MS_f associated to the flow $P\dot{V}$. The pseudo bonds ensuring the coupling between the pseudo elements introduced above, have been represented by dotted lines on the pseudo bond-graph of the chamber given in figure 2.71.

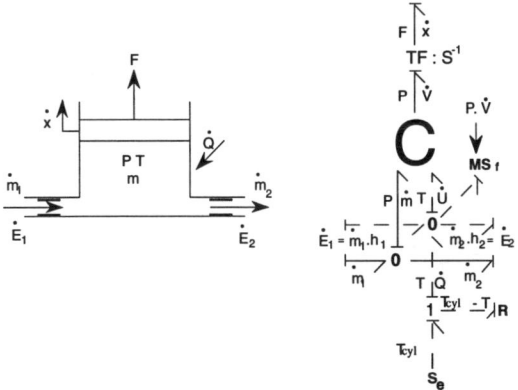

Figure 2.71: Pseudo bond-graph of a cylinder chamber.

2.4.3.2 Orifice

The flow of a compressible fluid through an orifice is represented by a four-port-pseudo R element, with a prescribed causality. This element allows the computation of the mass flow rate and the energy flow crossing an orifice in any flow direction and any flow type, subsonic or sonic. A computation procedure, given in the annex, is associated with the element shown in figure 2.72 in which can be found the two kinds of pseudo bonds previously defined. One will notice that this element is modulated by the term AC_q, product of the orifice section area A by the flow coefficient C_q. This crossing area varies with the spool position, i.e. with the virtual current previously introduced. Usually the flow coefficient is a function of the pressure ratio P_{down}/P_{up} and of the spool position and consequently of the virtual current i_v. Therefore, we propose globally taking into account the product AC_q as an experimental function of the pressure ratio between the downstream and upstream pressures and the virtual current i_v.

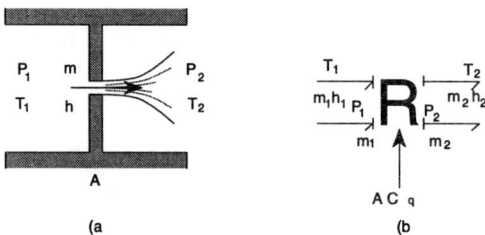

(a   (b

Figure 2.72: Orifices and associated pseudo bond-graph.

2.4.3.3   Flow stage- chamber assembly

The whole pneumatic part is composed of four orifices of the servovalve flow stage and of the two cylinder chambers.

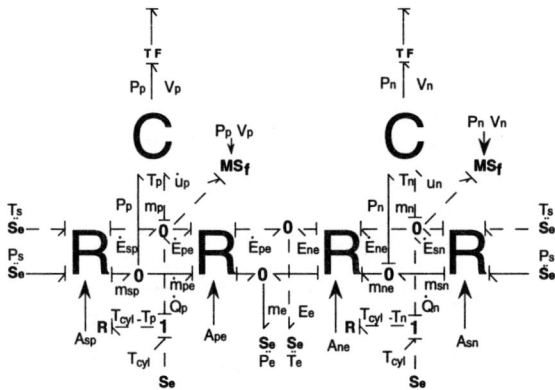

Figure 2.73: Pseudo bond-graph of the flow stage-chamber assembly.

The pseudo bond-graph of the flow stage is obtained (fig. 2.73):

- by associating with each orifice a four-port-pseudo R element;
- by introducing between the two pseudo R elements of a half-bridge, the pair of 0-junctions which are needed to take into account the mass balance and the energy balance in the chamber corresponding to the half-bridge under consideration;
- by introducing a pair of 0-junctions to connect the two R elements to a pseudo pressure source and a pseudo temperature source representing the atmosphere;
- by connecting each pseudo R element to a pseudo pressure source and a pseudo temperature source representing the supply.

By adding, to the flow stage pseudo bond-graph, two three-port pseudo C elements corresponding to the two chambers and two transformers, which enable the transformation of pneumatic energy into mechanical energy, into the chambers p and n, we obtain the pseudo bond-graph of the whole pneumatic part.

2.4.4 Bond-graph of the piston-load carriage assembly

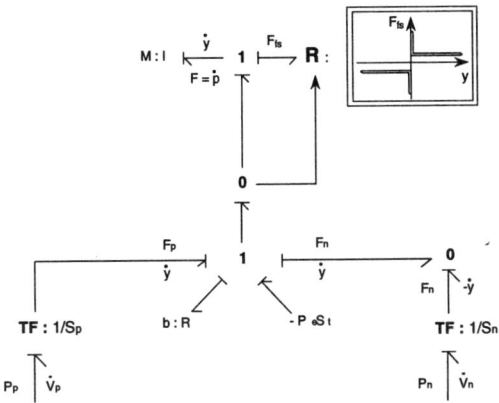

Figure 2.74: Bond-graph of the mechanical part.

Besides the two transformers previously mentioned, the bond-graph of the mechanical part (fig. 2.74) consists of the following elements, having the flow variable in common:

- a linear I-element taking into account the mass M of the mechanical part in translation,
- an effort source modeling the atmospheric pressure force against the cylinder rod,
- a linear R-element associated with viscous friction,
- a nonlinear R-element of prescribed causality representing dry friction in movement for the positive and negative values of the velocity. Moreover the constitutive relation for the initial start-up movement is different from the one associated with the following movements. This constitutive relation is experimentally defined from the experimental curve giving the dry friction as a velocity function. The two-bond 0-junctions enables us to show the sum of the efforts necessary for the implementation of the constitutive relation procedure of the nonlinear R-element. This variable modulates the R-element by an activated bond.

2.4.5 Bond-graph of the flow stage-actuator assembly

The bond-graph of the flow stage-actuator assembly is obtained by associating the bond-graph of the mechanical part and the pseudo bond-graph of the pneumatic part. On this bond-graph one can note the activated bonds which are necessary to take into account:

- the spool positioning dynamic by using a transfer function relating the current i to the virtual current i_v;
- the nonlinear dependence of the product AC_q, modulating the four-port pseudo R-element, upon the virtual current and the downstream and upstream pressures at the orifice.

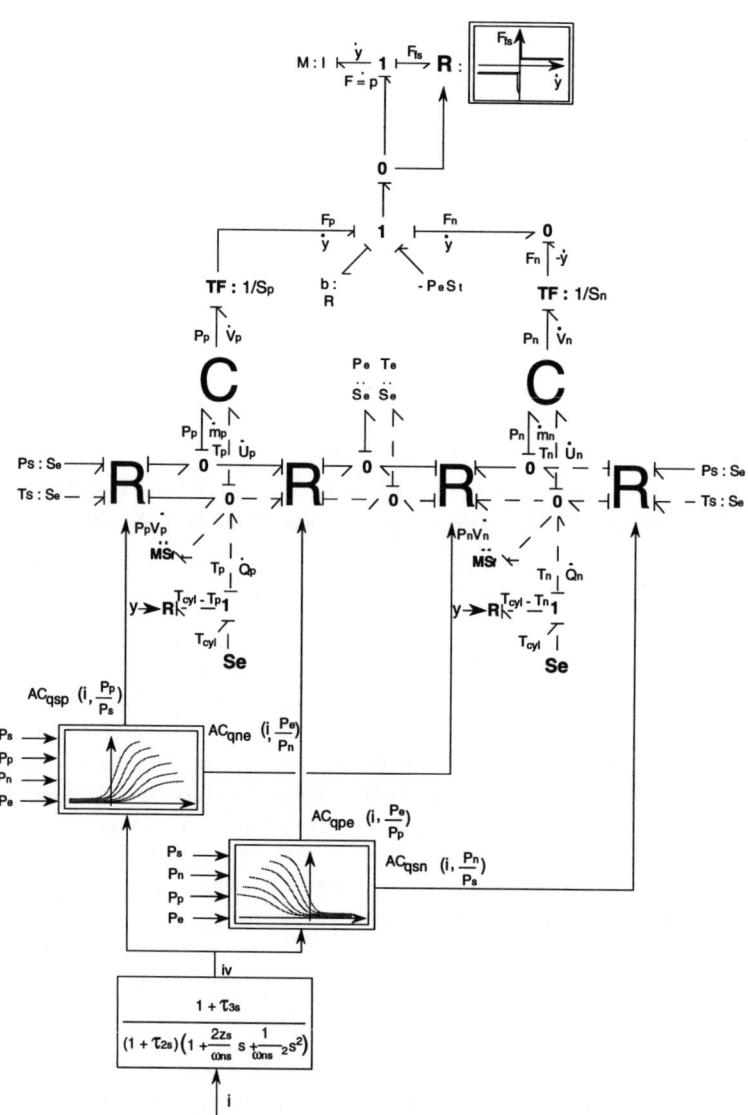

Figure 2.75: Pseudo bond-graph of the piston-load carriage assembly.

2.4.6   Bond-graph model for effort control

Passing from the position control to the effort control implies a model change because the effort control involves adding a stop. This is taken into account by adding a C element related to the 1-junction of the mechanical part and by suppressing the effort source related to the action of atmospheric pressure against the piston rod.

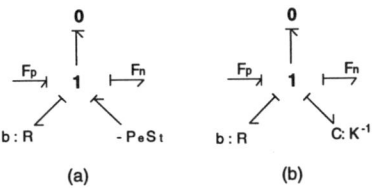

(a) position control model (no contact with the stop)
(b) effort control model (contact with the stop)

Figure 2.76: Bond-graph associated with the stop.

2.4.7   Bond-graph exploitation

2.4.7.1   Establishing the state equation (position control)

The causal bond-graph in figure 2.75 shows that the three storage elements, i.e. the two three-port pseudo C-elements and the I-element, have an integral causality on each of their bonds. To these three elements correspond seven energy variables susceptible to constituting state variables: U_p, m_p, V_p, U_n, m_n, V_n, p.

Moreover two direct causal paths can easily be seen on this bond-graph, the first between $\dot{y}$ and $\dot{V}_p$, the second between $\dot{y}$ and $\dot{V}_n$. The gain of these causal paths being equal to a constant, there is a proportionality between $\dot{V}_p$ and $\dot{V}_n$. After the integration there is the possibility of reducing to six the number of dynamically independent energy variables. From a bond-graph point of view we are led to choose the six following state variables: p, U_p, m_p, U_n, m_n, and either V_p or V_n.

It must be noticed that the position y appears explicitly as a state variable on the bond-graph only in the case where the actuator works against a spring (having a fixed extremity) or a stop (effort control).

Let us underline that, as there is an algebric relation between $\dot{y}$ and $\dot{V}_p$ (or $\dot{V}_n$), adding an energy storage term does not increase the number of state variables.

Then the state equation related to the generalized momentum has the expression:

$$\frac{dp}{dt} = \frac{rS_pU_p}{C_vV_p} - \frac{bp}{M} - \frac{rS_nU_n}{C_vV_n} - P_eS_t - F_{fs}. \tag{2.78}$$

The state equations associated with m and U are obtained by using the causal bond-graph of each chamber and of the related orifices.

For example, for the chamber p we can successively obtain:

• *State equation related to m_p*

$$\frac{dm_p}{dt} = \dot{m}_{sp} - \dot{m}_{pe} \quad \text{(pseudo 0-junction associated with the chamber } p\text{)},$$

with:

$$\dot{m}_{sp} = AC_{qsq}.C_m\left(\frac{P_p}{P_s}\right).\frac{P_s}{(T_s)^{1/2}} \quad \text{(pseudo } R\text{-element associated with the orifice } A_{sp}\text{)},$$

$$\dot{m}_{pe} = AC_{qpe}.C_m\left(\frac{P_e}{P_p}\right).\frac{P_p}{(T_p)^{1/2}} \quad \text{(pseudo } R\text{-element associated with the orifice } A_{pe}\text{)},$$

(2.79)

$$\left.\begin{array}{l} P_p = \dfrac{r.U_p}{C_v.V_p} \\[2mm] T_p = \dfrac{U_p}{C_v.m_p} \end{array}\right\} \quad \begin{array}{l} \text{constitutive relations of the pseudo } C\text{-element} \\ \text{associated with the chamber } p. \end{array}$$

• *State equation related to U_p*

$$\frac{dU_p}{dt} = \dot{E}_{sp} - \dot{E}_{pe} + \dot{Q}_p - P_p\dot{V}_p,$$

$$\dot{E}_{sp} = \dot{m}_{sp}.C_p.T_s \quad \text{(pseudo } R\text{-element associated with the orifice } A_{sp}\text{)},$$

$$\dot{E}_{pe} = \dot{m}_{pe}.C_p.T_p \quad \text{(pseudo } R\text{-element associated with the orifice } A_{pe}\text{)},$$

(2.80)

$$\dot{Q}_p = kA_p(y)(T_{cyl} - T) \quad \text{(pseudo } R\text{-element in the thermal domain)}.$$

A similar approach leads to the state equations related to the chamber n.

• *State equation related to m_n*

$$\frac{dm_n}{dt} = \dot{m}_{sn} - \dot{m}_{ne},$$

with:

$$\dot{m}_{sn} = AC_{qsn}.C_m\left(\frac{P_n}{P_s}\right).\frac{P_s}{(T_s)^{1/2}} \quad \text{and} \quad P_n = \frac{r.U_n}{C_v.V_n},$$

$$\dot{m}_{ne} = AC_{qne}.C_m\left(\frac{P_e}{P_n}\right).\frac{P_n}{(T_n)^{1/2}} \quad \text{and} \quad T_n = \frac{r.U_n}{C_v.m_n}.$$

(2.81)

• *State equation related to U_n*

$$\frac{dU_n}{dt} = \dot{E}_{sn} - \dot{E}_{ne} + \dot{Q}_n - P_n\dot{V}_n,$$

with:

$$\begin{array}{l} \dot{E}_{sn} = \dot{m}_{sn}.C_p.T_s \\[2mm] \dot{E}_{ne} = \dot{m}_{ne}.C_p.T_n \end{array} \quad \text{and} \quad \dot{Q}_n = k.A_n(y).(T_{cyl} - T_n).$$

• *State equation related to* V_p

$(\mathrm{d}V_p)/(\mathrm{d}t) = S_p p/M$ (gain of the causal path between $\dot{V}_p$ and the coenergy variable $\dot{y}$ of the I-element and constitutive relation of this element).

• *Algebraic equations*

$$V_n = V_{mn} + \frac{cS_n}{2} - \frac{S_n}{S_p}\left(V_p - V_{mp} - \frac{S_p c}{2}\right),$$

$$y = \frac{1}{S_p}\left(V_p - V_{mp} - \frac{S_p c}{2}\right).$$

(2.82)

2.4.7.2 Display of the couplings

By taking into account the coupling admitted by the multiport elements, the study of causal paths shows:

- for each chamber a causal loop between the pseudo C-element and the pseudo R-element (associated with the thermal exchange by convection) shows a "purely thermal dynamic";
- for each half-bridge and associated chamber, three causal loops between the pseudo C-element and the two pseudo R-elements of the half-bridge under consideration. These loops show the influence of the orifices on the chamber pressure and temperature changes;
- for the piston-load carriage assembly: a causal loop between the I-element and the two R-elements associated with various friction types. All these loops must be respectively related to the cylinder natural pulsation and to the associated damping.

Calculating the time constants, the natural pulsations and the reduced damping coefficients corresponding to these loops, for the linearized tangent system, implies making the bond-graph of this linearized tangent system.

By using the initial bond-graph one can study the influence of certain modeling hypothesis. For example the presence of leakage between the cylinder chambers leads to the addition of a four-port pseudo R-element between the two pseudo C-elements and make new causal loops appear.

2.4.8 Conclusion

Using bond-graphs for the representation of this type of system is interesting because there is:

- a unified approach to the different domains (the establishment of the state model of the whole servovalve, including the electromechanical part, would have made this argument stronger);
- a graphical display of coupling between the different domains that leads to better understanding of the system behavior;
- a facility for modifying the model when adding or suppressing certain elements without modifying the physical junction structure;
- a possibility of associating several types of representation (transfer function, bond-graph, pseudo bond-graph) to help development of hypothesis by the analyst.

Notation

A: cross-sectional area of the orifice (m^2);

b: viscous friction coefficient (N/ms^{-1});

c: piston stroke (m);

C: capacitor;

C_q: flow coefficient;

C_p: specific heat at constant pressure $(Jkg^{-1}K^{-1})$;

C_m: flow parameter $(m^{-1}\sqrt{K}s)$;

C_v: specific heat at constant volume $(Jkg^{-1}K^{-1})$;

E: energy (J);

$\dot{E}$: energy flux (W);

F: force (N);

F_v: force due to viscous friction (N);

F_{fs}: force due to dry friction (N);

h: specific enthalpy (J/kg);

i: servovalve current (A);

i_v: virtual current (A);

k: convection coefficient $(Jm^{-2}K^{-1})$;

m: mass flow rate contained in a chamber (kg);

$\dot{m}$: mass flow rate (kgs^{-1});

M: mass of the moving system (kg);

P: absolute pressure (Pa);

P_r: ratio between downstream and upstream pressures;

p: generalized momentum $(kgms^{-1})$;

q: capacitor charge;

$\dot{Q}$: exchanged flux by convection (w);

r: perfect gas constant related to the unit mass $(Jkg^{-1}K^{-1})$;

S: piston area (m^2);

T: absolute temperature (K);

T_{cyl}: cylinder temperature (K);

U: internal energy of the gas contained in a chamber (J);

V: chamber volume (m^3);

V_{mi}: dead volume of a chamber i (m^3);

v: velocity (ms^{-1});

x: displacement (m);

y: piston displacement (m);

z: servovalve spool position (m).

Indices

up: variable related to the upstream of the orifice;

dow: variable related to the downstream of the orifice;

e: variable related to the exhaust;

i: variable related to the chamber *i*;

ij: variable related to an orifice of upstream pressure P_i and downstream pressure P_j;

o: initial value of a variable;

s: variable related to the supply;

1: input variable for a chamber;

2: output variable for a chamber;

γ: ratio of specific heat at constant pressure and constant volume.

Annex: Procedure associated with the pseudo *R*-element shown in figure 2.72 (b)

On this four-port pseudo *R*-element the causality is fixed. The pressures P_1, P_2 and the temperatures T_1 and T_2 are given, the mass flow rates $\dot{m}_1$, $\dot{m}_2$ and the enthalpy flows $\dot{m}_1 h_1$ and $\dot{m}_2 h_2$ are calculated according to the following procedure:

1. Detection of downstream and upstream pressures:

   If $P_1 > P_2$ then $P_{up} = P_1$, $T_{up} = T_1$, $P_{down} = P_2$.

   If $P_2 \geq P_1$ then $P_{up} = P_2$, $T_{up} = T_2$, $P_{down} = P_1$.

2. Computation of the pressure ratio P_r:

   If $P_{down}/P_{up} > P_{cr}$ then $P_r = P_{down}/P_{up}$.

   If $P_{down}/P_{up} \leq P_{cr}$ then $P_r = P_{cr}$,

   with:

   $$P_{cr} = \left(\frac{2}{\gamma+1}\right)^{\frac{\gamma}{\gamma-1}}.$$

3. Computation of the absolute value of the mass flow rate:

   $$\dot{m} = \frac{AC_q P_{up}}{(T_{up})^{1/2}} C_m \left(\frac{P_{down}}{P_{up}}\right),$$

   with:

   $$P \geq P_{cr} \quad C_m = \left(\frac{2\gamma}{r(\gamma-1)}\right)^{1/2} \left[P_r^{\frac{2}{\gamma}} - P_r^{\frac{\gamma+1}{\gamma}}\right]^{1/2},$$

   $$P \leq P_{cr} \quad C_m = \left(\frac{\gamma}{r}\left(\frac{2}{\gamma+1}\right)^{\frac{\gamma+1}{\gamma-1}}\right)^{\frac{1}{2}}$$

   r gas perfect constant.

4. Sign assignment as a function of the flow direction:

   If $P_1 > P_2$ $\dot{m}_1 = \dot{m}_2 = \dot{m}.$

   If $P_1 \leq P_2$ $\dot{m}_1 = \dot{m}_2 = -\dot{m}.$

5. Computation of the enthalpy flow rate:

$$\dot{m}_1 h_1 = \dot{m}_2 h_2 = \dot{m}_1.C_p.T_{up},$$

C_p specific heat at constant pressure.

2.5   Bibliography

[1] BORNE P., DAUPHIN-TANGUY G., RICHARD J.P., ROTELLA F., ZAMBET-TAKIS I. (1992), *"Modélisation et Identification des processus"*, ch. 5, vol. 2, Collection "Méthodes et pratiques de l'ingénieur", Edition Technip.

[2] BOS A.M. (1986), *"Modeling multibody systems in terms of multi bond-graphs with application to a motorcycle"*, PhD thesis., University of Enschede, the Netherlands.

[3] BREEDVELD P. (1984), "Essential gyrators and equivalence rules for 3-port junction structure", *J. of the Franklin Inst.,* 318, n.2, pp.77–89.

[4] BREEDVELD P.C. (1984), *"Physical systems theory in terms of bond-graphs"*, PhD thesis. Twente Un. of Technology Enschede, The Netherlands, ISBN-Ní90-9000599-4, 201 p.

[5] BROWN F.T. (1972), "Direct application of the loop rule to bond-graphs", *J. Dyn. Syst., Meas. and Cont.,* 3, pp. 253–261.

[6] BUYSE H. (1984), *"Electrical machines and converters modeling and simulation"*, North Holland.

[7] DAUPHIN-TANGUY G., BORNE, P., LEBRUN, M. (1985), "Order reduction of multi-time scale systems using bond-graphs, the reciprocal system and the singular perturbation method", *J. of the Franklin Institute,* 319, pp. 157–171.

[8] DET F., SCAVARDA S., RICHARD E. (1988), *Modeling and simulation of a position and force control of an electropneumatic gripper with bond-graphs,* In: Proc. 12th IMACS World Congress, Paris, July 18–22, 18 p.

[9] DUCREUX J.P., CASTELAIN A., DAUPHIN-TANGUY G., ROMBAUT C. (1992), *"Power electronics and electrical machines modeling using bond-graphs"*, IMACS Transaction on "Bond-graphs for Engineers", North Holland-Elsevier, eds G. DAUPHIN-TANGUY, P. BREEDVELD, pp. 121–134.

[10] KARNOPP D.C. (september 1979), State variables and pseudo bond-graphs for compressible thermofluid systems, *J. of Dynamic Systems Measurement and Control*, vol. 101, pp. 201–204.

[11] KARNOPP D.C., MARGOLIS D.L., ROSENBERG R.C. (1990), *Systems dynamics: a unified approach*, John Wiley and Sons, (2nd edition).

[12] LEBRUN M., SCAVARDA S. (1979), Simulation of the nonlinear behavior of an electrohydraulic exciter, *Simulation*, vol. 33, N14, pp. 127–141.

[13] LE DŒUFF R. (1989), *"Enseignement de la simulation numérique en électronique de puissance"*, Journées EEA sur la "CAO et simulation électronique".

[14] MANESSE G. (1987), *"Sur une analyse fonctionnelle de groupements d'interrupteurs statiques - Extension à la modélisation des convertisseurs dans leur environnement de contrôle"*, PhD., Lille.

[15] MASCHKE B. (1990) *"Contribution à une approche par bond-graph de l'étude et la conception de lois de commande de robots contenant des segments flexibles"*, Doctorate thesis, Université d'Orsay-Paris Sud.

[16] PAYNTER M. (1961), *Analysis and design of engineering systems*, MIT Press.

[17] ROSENBERG C., KARNOPP D. (1983), *Introduction to physical system dynamics*, Mechanical Engineering Series, McGraw Hill.

[18] SUEUR C., DAUPHIN-TANGUY G. (1991), "Bond-graph approach for structural analysis of MIMO linear systems", *J. of the Franklin Institute,* 328, n.1, pp. 55–70.

[19] SUEUR C., DAUPHIN-TANGUY G. (1991), "Bond-graph approach for multi-time scale systems analysis", *J. of the Franklin Institute,* special issue on "Current topics in bond-graph related research", vol. 328, n. 5/6, pp. 1005–1026.

[20] THOMA J. (1991), *Simulation by bond-graphs*, Springer Verlag.

CHAPTER 3

Identifiabilities and nonlinearities

E. WALTER, L. PRONZATO

Abstract: A parametric model structure is (globally) identifiable if the parameter vector associated with a given input-output behavior is unique. If this is not the case, careless estimation of this vector from experimental data may lead to completely erroneous results. (The situation is similar when unobserved state variables have to be estimated from the outputs.) The effect of two types of nonlinearity of the output (with respect to the parameters and to the inputs) on structural identifiability is described. Various methods to test state space models for identifiability are presented. They apply to models that are nonlinear with respect to the parameters and may be linear or not with respect to the inputs. Their use is illustrated on simple examples and an actual problem in chemical engineering. Advantages and limitations of the existing techniques are evidenced. Relationships between structural identifiability and practical identifiability, i.e. the ability to actually estimate the parameters of the model from experimental data, are considered, as well as techniques available to design an experiment so as to make the parameters of interest as identifiable as possible.

3.1 Modeling and parameter estimation

Before addressing the notion of identifiability and its application to nonlinear models more in detail, it might be useful to put them into their context, which is building models from experimental data. This operation is at the heart of any scientific proceedings, and an overview is given here. For more details, one could for instance refer to Ljung (1987), Norton (1986), Walter and Pronzato (1994).

Building a mathematical model to describe the behavior of a process may serve many purposes (understanding of a physical phenomenon, simulation for teaching, prediction in order to design a suitable control law, estimation of quantities that are not directly measurable, fault detection etc.). Most often, prior knowledge is not sufficient to allow a complete derivation of the model, and its equations involve unknown parameters. These parameters form a vector p to be estimated from the experimental data collected on the process. Let $M(.)$ be the structure chosen for the model (e.g., a first-order linear differential equation with constant coefficients) and $M(p)$ be the specific model obtained when the parameters take the value p. The choice of $M(.)$ defines the class of possible models and is a critical matter, since one can only hope obtaining the best possible model in the class considered. An unsuitable choice may therefore result into useless attempts at estimating the corresponding parameters. In many cases, the structure to be selected for the model cannot be deduced entirely from prior knowledge, and a set of possible structures $\{M_i(.), i = 1, \cdots, n_s\}$ must be considered, with which are associated as many vectors p_i of unknown parameters. One may for instance consider all linear differential equations with constant coefficients of order between 1 and 5.

One would then like to use the data collected on the process to select the most appropriate structure and estimate its parameters. This corresponds to the connected problems of structure discrimination and parameter estimation. The usual procedure consists in estimating at best (in a sense to be specified) the parameters of each, structure starting with the simplest ones (e.g., those with the smallest number of unknown parameters). The simplest structure leading to a satisfactory behavior is then selected.

In order to compare models, one needs a scale of values, or criterion, allowing them to be ranked. The value of the criterion associated with the model $M(p)$ will depend on the data collected on the process $\{y(t_i), i = 1, \ldots, n_t\}$ and on the values of the corresponding quantities computed by simulation of $M(p)$, which will be denoted by $\{y_m(t_i, p), i = 1, \ldots, n_t\}$. The outputs y and y_m may depend on inputs applied to the process, even if this dependency is not made explicit here. Note that the linear or nonlinear character of the output y_m of the model with respect to its inputs has relatively little influence on the nature of the estimation problem (it might only require changing the simulation algorithm to be used). On the other hand, the linear or nonlinear character of y_m with respect to the parameters p will have a large impact. We shall return to this later.

The most frequently used criterion is quadratic and defined by

$$J(p) = \sum_{i=1}^{n_t} [y(t_i) - y_m(t_i, p)]^T Q_i [y(t_i) - y_m(t_i, p)],$$

where the weighting matrices Q_i (symmetric, non-negative definite) quantify the cost to be paid for incorrectly reproducing $y(t_i)$.

Various statistical approaches can be used to build criteria in a rational manner, based on some properties that the model should have. If, for instance, one wishes the error $y(t_i) - y_m(t_i, p)$ to correspond to realizations of independent random variables identically normally distributed with zero mean and covariance $\sigma^2 I$ (where I is an identity matrix), *the maximum-likelihood approach* leads to minimizing a quadratic criterion with $Q_i \equiv I$. The maximum-likelihood estimator possesses interesting asymptotic properties (convergence, normality, efficiency). *Bayesian techniques*, based on the knowledge of a prior probability density function for the parameters, allow the final goal of the modeling to be taken into account via a quantification of the cost of an erroneous choice for the parameters. The choice of this cost function also has a critical influence on the definition of the best model.

Once the structure $M(.)$ has been chosen and the criterion $J(.)$ has been defined, one must search for the value $\hat{p}$ of the parameter vector that is optimal in the sense of $J(.)$. The usual procedure corresponds to a scheme similar to that described in figure 3.1. The system S is only accessible via its mastered inputs u, and its outputs y. The vector b stands for the perturbations (unmastered inputs, measurement errors, approximative nature of the model). The estimates of the state variables (or of internal variables) of the system that cannot be measured directly are generated by the model $M(p)$ and denoted by z_m.

For any considered model structure, the parameters are modified in order to optimize $J(.)$. The structure finally selected is the one that turns out to be the best in the sense of the criterion (possibly weighted by a term penalizing the most complex structures).

The resulting optimization criteria have some interesting common features.

(i)   The number of parameters is small (typically less than 10).

(ii) The analytical calculation of the gradient and Hessian of the criterion $\left(\dfrac{\partial J}{\partial p}\right.$ and $\left.\dfrac{\partial^2 J}{\partial p \partial p^T}\right)$ is generally possible.

(iii) The problem is often unconstrained, any possible constraint on the feasible values for the parameters being used *a posteriori* to check the validity of the results obtained.

(iv) The problem is frequently ill-conditioned, for the influence of the various parameters on the value of $J(.)$ may differ by several orders of magnitude.

(v)   The problem is often not convex and several local or even global optima may exist.

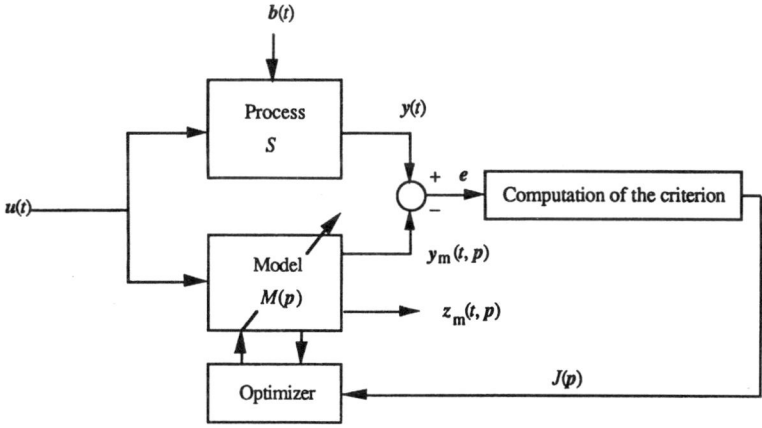

Figure 3.1: Estimation scheme.

The techniques to be implemented to find $\hat{p}$ depend on the nature of the dependency of $J(.)$ on p. If $J(.)$ is quadratic in p (for instance if it corresponds to the quadratic criterion defined above and if y_m depends on p linearly), then $\hat{p}$ can be obtained explicitly by the *least-squares method,* and the unicity of the optimum is guaranteed, provided some (easy to check) identifiability conditions are satisfied. Various methods (*extended least squares, generalized least squares, extended matrix, instrumental variables, etc.*) make it possible to extend the application of the least-squares approach to more general situations. Another approach takes advantage of the local properties of the criterion in the neighborhood of the current point $\hat{\boldsymbol{p}}^k$ in the parameter space in order to deduce a new point $\hat{\boldsymbol{p}}^{k+1}$ in such a way as to improve $J(.)$. Various methods guarantee convergence towards a local optimum of $J(.)$ that may depend on the initial point $\hat{\boldsymbol{p}}^0$. The Gauss-Newton algorithm (and its variant due to Levenberg and Marquardt), the quasi-Newton algorithms (e.g. the one by Davidon, Fletcher and Powell) and the conjugate gradient algorithms are among the most popular and are included in most scientific subroutine libraries. They use the value of the criterion and its gradient (or sensitivity functions) at the current point. The gradient can be approximated by finite differences or computed rigorously with adjoint-state techniques, while the sensitivity functions of the model output with respect to the parameters can be obtained by simulating equations deduced from those defining the model. Whatever the method chosen, these quantities often have to be computed a great number of times through the optimization, which will involve a great number of simulations. It is therefore important to have powerful simulation tools at one's disposal.

All these techniques are essentially local and cannot in general guarantee convergence towards a global optimum of the criterion, even if the optimization is started from a large number of initial vectors $\hat{\boldsymbol{p}}^0$ drawn at random in the prior feasible space. Let us mention, however, the existence of global optimization algorithms that make it possible to locate all global optima of the criterion as precisely as desired (Ratschek and Rokne, 1988; Horst and Tuy, 1990). Because of the intensive nature of the required computations, these algorithms can only be used on problems with a limited number of parameters.

Once the best value for the parameters of $M(.)$ in the sense of $J(.)$ has been obtained, various techniques make it possible to characterize the uncertainty in these parameter estimates. Some of these techniques rely on intensive computations (Monte-Carlo methods, bootstrap, etc.). A much simpler approach takes advantage of the asymptotic properties of maximum-likelihood estimators and assimilates the covariance matrix of the estimation error to the inverse of the Fisher information matrix, which is easily computed from the hypotheses on the noise statistics already used to build the estimator. This method is by far the most popular, because it involves almost no additional computations. The very approximate nature of the results thus obtained when there are few data points must however be stressed.

The Fisher information matrix depends on the experimental conditions under which the data have been collected (measurement times, input shape, position and quality of the sensors and actuators, etc.). Advantage can be taken of this property to optimize data collection (*experiment design*). Among all feasible experiments, one will then look for the one that can be expected to yield the most pertinent information. One may, for instance, maximize the determinant of the Fisher information matrix, which corresponds to the notion of D-optimal design and amounts to minimizing the volume of asymptotic confidence ellipsoids for the parameters. We shall return to the notion of experiment design and its connexions with identifiability in section 3.7.

The various steps in model building (choice of $M(.)$, of $J(.)$, of the optimization algorithm, initialization, etc.) rely on hypotheses that are not always explicit and are more or less justified. It is therefore important to keep a critical eye on the results obtained and to be prepared to question them.

In this sense, it will be useful to submit the model finally obtained to a set of trials attempting to invalidate (or falsify) it. One may for instance examine the quality of the predictions provided by the model on a data set that differs from that used to estimate the parameters.

The first of these questions may be performed during the choice of the model structures to be considered. Analysing the properties of these structures, and especially their identifiability and distinguishability, enables detection of serious defects even before any experimental data have been collected.

The next example will illustrate a type of problems for which these notions are of great practical importance.

Chemical engineering example The experimental set-up described in figure 3.2 is used in the Department of Chemical Engineering and Applied Chemistry of Columbia University in New York to study, with the help of isotopic transients, the reaction that produces methane from carbon monoxide, according to

$$CO + 3H_2 \iff CH_4 + H_2O.$$

Before the initial time, a gaseous mixture of carbon monoxide, methane, hydrogen and water is circulated at high speed on a solid nickel-based catalyst. Consumed carbon monoxide and hydrogen are permanently replaced with the help of a syringe, while pressure and temperature are kept constant by appropriate regulations evacuating a corresponding quantity of gas at the outlet of the reactor. The composition of the gaseous mixture thus collected is analyzed with a mass spectrometer. If one waits long enough, the reactor reaches a stationary state, and the composition of the gaseous mixture becomes approximately constant. This corresponds to a dynamic equilibrium, where each molecule leaving the reactor is replaced by an identical one. At time $t = 0$, the syringe containing ^{12}CO is replaced by one containing ^{13}CO. The evolution with time of the percentage of marked carbon atoms in the carbon monoxide and methane leaving the reactor can then be followed. Assuming that the reactor is in a stationary state and that there is no isotopic effect (i.e. the atoms of ^{13}C and ^{12}C behave identically with respect to the reaction), the evolution of the tracer is linear (and time-invariant) with respect to the input.

Chemical prior information leads to two structures competing for the description of the isotopic data (figures 3.3 and 3.4). These structures follow the formalism of compartmental models, frequently used in biology to analyse the results of experiments based on isotopic tracers. Compartments, represented by squares, correspond to the various chemical species that can be labeled by the tracer. The exchanges of labelable atoms between compartments are materialized by arrows on which the flows of labelable atoms are indicated. In both structures, the state variable x_i is the specific activity (percentage of labeled atoms) in compartment i. The physical meaning of the various compartments on figures 3.3 and 3.4 is as follows:

- compartment 1 = active carbon on the catalyst surface
- compartment 2 = inactive carbon on the catalyst surface
- compartment 3 = adsorbed hydrocarbons
- compartment 4 = gaseous methane.

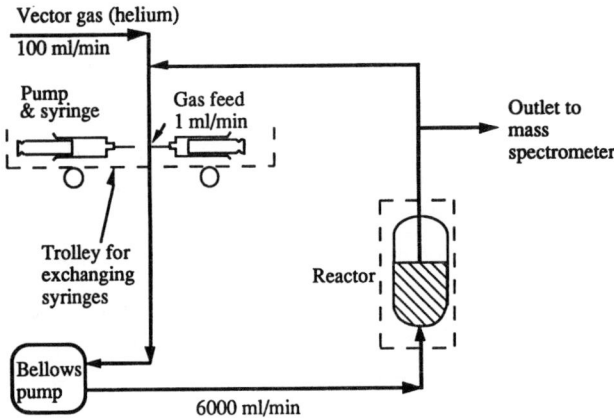

Figure 3.2: Experimental set-up in heterogeneous catalysis.

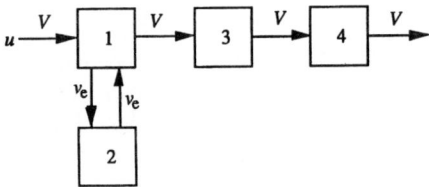

Figure 3.3: First model structure.

State equations express the balance of the exchanges of atoms between the compartments of the model. The first structure is described by the set of equations

$$M(p) : \begin{cases} C_1 \dfrac{d}{dt} x_1 &= -(V + v_e)x_1 + v_e x_2 + Vu, \\[2mm] C_2 \dfrac{d}{dt} x_2 &= v_e x_1 - v_e x_2, \\[2mm] C_3 \dfrac{d}{dt} x_3 &= Vx_1 - Vx_3, \\[2mm] C_4 \dfrac{d}{dt} x_4 &= Vx_3 - Vx_4, \\[2mm] y_m &= x_4, \end{cases}$$

with $x(0) = 0$. The parameters to be estimated are

$$p = (C_1, C_2, C_3, v_e)^T.$$

The $C_i (i = 1, 2, 3)$ are surface concentrations, and v_e is the velocity of the transfer of carbon atoms between adsorbed species (from compartment 1 to compartment 2 and from compartment 2 to compartment 1). The quantities V (overall velocity of the transfer of carbon atoms resulting from the reaction) and C_4 (volume fraction of methane in the gaseous

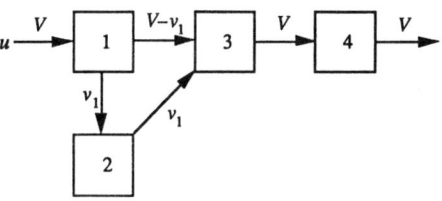

Figure 3.4: Second model structure.

phase) are known from independent measurements. The input u is the percentage of labeled carbon atoms in the feed of the reactor.

The second model structure considered is described by the set of equations:

$$\hat{M}(\hat{p}) : \begin{cases} C_1\dfrac{d}{dt}x_1 = -Vx_1 + Vu, \\[2mm] C_2\dfrac{d}{dt}x_2 = v_1x_1 - v_1x_2, \\[2mm] C_3\dfrac{d}{dt}x_3 = (V - v_1)x_1 + v_1x_2 - Vx_3, \\[2mm] C_4\dfrac{d}{dt}x_4 = Vx_3 - Vx_4, \\[2mm] \hat{y}_m = x_4, \end{cases}$$

with $x(0) = 0$. The parameters to be estimated are

$$\hat{p} = (C_1, C_2, C_3, v_1)^T.$$

The $C_i (i = 1, 2, 3)$, V, u and C_4 are defined as previously; v_1 is the velocity of the transfer of carbon atoms between adsorbed species (from compartment 1 to compartment 2 and from compartment 2 to compartment 3).

The model structures considered here have some specific characteristics that are the hallmark of problems for which testing models for identifiability is of importance:

- state variables and parameters have a concrete meaning;
- the aim of the modeling is not limited to the satisfactory reproduction of an experimentally observed input-output behavior.

One has to select the more suitable model structure (discrimination or hypotheses testing) and to determine the numerical value of the associated parameters (parameter estimation). This is a problem of system analysis and modeling, and not of control of the output variables of a process. One thus hopes to be able to detect the characteristics of the slow step of the reaction and to use this information in order to improve the catalyst.

Before embarking on data collection, it is natural to ask whether there is any chance of succeeding, which can be summarized by the following questions:

- Will it be possible to decide from the data which structure is the right one (if any)?
- This being done, will it be possible to estimate its parameters in a meaningful way from the data?

The notions of structural identifiability and distinguishability presented in section 3.2 will make it possible to answer these questions in an idealized framework. When the answers turn out to be negative, they will also help to suggest remedies for the defects thus detected. As often happens in automatic control theory or statistics, the techniques to be used will depend on whether the model is linear or not. In section 3.3, two types of linearity of the model output will be distinguished, namely with respect to the inputs (LI) and with respect to the parameters (LP). Non-LP model structures are the only ones that present any difficulty, and the remainder of this chapter will be entirely devoted to them. Section 3.4 will recall the methods most commonly used to test LI models for structural identifiability and distinguishability. Section 3.5 will be devoted to non-LI models. Each method will be applied to a simple example treated in details. The links between existing techniques for LI and non-LI models will be evidenced. All the methods described require algebraic manipulations (especially elimination of variables) that may turn out to be very complex when one goes beyond academic examples. This is why it is necessary to make them systematic and implementable on computers, as we shall see in section 3.6, where the identifiability and distinguishability of the two compartmental model structures considered above will be studied. Finally, section 3.7 will connect the notion of structural identifiability with that of experiment design for parameter estimation, which can be seen as an attempt at maximizing practical identifiability.

3.2   Structural identifiability and distinguishability

3.2.1   Structural properties

Once a structure has been chosen for the model (or a set of structures among which one will have to choose), one needs to be able to study its properties as independently as possible of the numerical values taken by the parameters. As a matter of fact, the study of these properties should be allowed to take place even before the data have been collected. Let P be the prior feasible parameter space. A property will be said to be structural if it holds true for almost any value of p belonging to P, and possibly false on a subspace with zero measure. Thus a property that will be true for any value of p that does not belong to a hypersurface of the parameter space will be considered as structural, since the probability of drawing an atypical value of the parameters when selecting a point at random in P is zero. Two structural properties are of special importance in the context of modeling, namely *identifiability* and *distinguishability*.

3.2.2   Structural identifiability

We shall set ourselves in an idealized framework (figure 3.5) where

(i) the process and its model have the same structure $M(.)$ (no characterization error), and there is a true value p^* for the process parameters,

(ii) the data are noise-free ($b \equiv 0$),

(iii) the input applied to the process and measurement times can be chosen freely (and the input is computed independently of the output).

Under these conditions, it is always possible to tune the parameters of the model so as to give it the same input/output behavior as the process for any input, which we shall denote by

$$M(p^*) = M(\hat{p}).$$

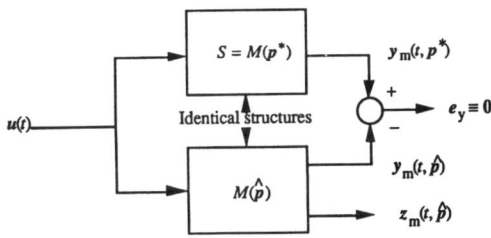

Figure 3.5: Idealized framework for structural identifiability studies.

We would like to know whether this identity of all possible input/output behaviors implies the equality of the parameters $\hat{p}$ of the "model" and p^* of the "process". More precisely, the *parameter p_i* is said to be *structurally globally identifiable (s.g.i.)* if

$$\text{for almost any } p^* \in P, M(\hat{p}) = M(p^*) \Longrightarrow \hat{p}_i = p_i^*.$$

We shall later see a concrete example of why the restriction to almost any p^* is necessary.

The *model $M(.)$* is *s.g.i.* if all of its parameters are s.g.i.

When one fails to prove that a model is s.g.i., one may be interested by the weaker notion of local identifiability. The *parameter p_i* is said to be *structurally locally identifiable (s.l.i.)* if for almost any $p^* \in P$, there exists a neighborhood $v(p^*)$ such that

$$\hat{p} \in v(p^*) \text{ and } M(\hat{p}) = M(p^*) \Longrightarrow \hat{p}_i = p_i^*.$$

Local identifiability is therefore a necessary condition for global identifiability.

The *model $M(.)$* is *s.l.i.* if all of its parameters are s.l.i.

The *parameter p_i* is said to be *structurally nonidentifiable (s.n.i.)* if for almost any $p^* \in P$, there is no neighborhood $v(p^*)$ such that

$$\hat{p} \in v(p^*) \text{ and } M(\hat{p}) = M(p^*) \Longrightarrow \hat{p}_i = p_i^*.$$

The *model $M(.)$* is *s.n.i.* if at least one of its parameters is s.n.i. Note that some parameters of an s.n.i. model may very well be s.l.i. or even s.g.i. A model with serious identifiability defects may therefore still present some interest since it might allow the estimation of some parameters or the determination of all posterior feasible values of others. Note finally that a structural conclusion cannot always be reached, because some parameters may be neither s.n.i. nor s.l.i. The conclusion will then depend on the numerical value of the parameters.

Remark 1

If the vector p of the parameters is included in an extended state vector, the techniques available to test the structural observability of this extended state can be used to study identifiability (Åström, 1972; Berntsen and Balchen, 1973; Diop and Fliess, 1991).

3.2.3   Structural distinguishability

One often has to choose among several candidate structures. It is then natural to wonder whether the measurements to be performed on the process will make it possible to select the most suitable of them. This question of structural *distinguishability* receives a partial answer in the same idealized framework as that of structural identifiability. One therefore assumes (figure 3.6) that the "process" is a model with structure $M(.)$ while the "model" is a model with a structure $\hat{M}(.)$ that differs from $M(.)$. The parameters of $\hat{M}(.)$ will be denoted by $\hat{p}$, and those of $M(.)$ by p^*. These vectors are not necessarily of the same dimension. Since the "process" and its "model" no longer have the same structure, it is no longer evident that it will be possible to tune the parameters $\hat{p}$ of the "model" so as to obtain the same input/output behavior as for the "process".

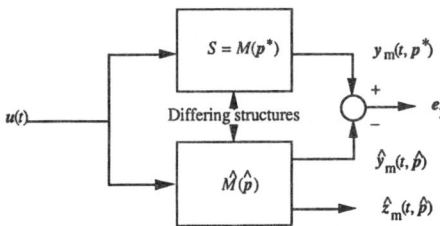

Figure 3.6: Idealized framework for structural distinguishability studies.

On the contrary, one hopes that it will be possible to notice different behaviors. More precisely, $\hat{M}(.)$ will be said to be *structurally distinguishable (s.d.)* from $M(.)$ if for almost any feasible value p^* of the parameters of $M(.)$ there is no feasible value $\hat{p}$ of the parameters of $\hat{M}(.)$ such that $\hat{M}(\hat{p}) = M(p^*)$.

Note the asymmetry of the previous definition. The fact that $\hat{M}(.)$ is s.d. from $M(.)$ does not imply that the converse is true. One class of models may for instance be a subset of the other (although this might not necessarily be obvious at first sight). When $\hat{M}(.)$ is s.d. from $M(.)$ and $M(.)$ is s.d. from $\hat{M}(.)$, $M(.)$ and $\hat{M}(.)$ will be said to be s.d.

With the help of counterexamples (Walter, Lecourtier and Happel, 1984), it is easy to prove that the identifiability of two structures is neither necessary nor sufficient for them to be distinguishable. The techniques to be used to test these two types of properties are nevertheless similar. In what follows, we shall stress methods for testing structural identifiability, from which methods for testing distinguishability are trivial to deduce. We shall need to distinguish two types of nonlinearities.

3.3 Nonlinearities

Let $y_m(t, p, u)$ be the output of the model at time t when the input $u(.)$ has been applied between times 0 and t to a model in zero initial condition at $t = 0$. A model is *nonlinear with respect to its inputs (non-LI)* if its output does not satisfy the superposition principle with respect to its inputs, i.e. if

$$\exists \lambda, \mu \in \mathbb{R}, t \in \mathbb{R}^+ |\; y_m(t, p, \lambda u_1 + \mu u_2) \neq \lambda y_m(t, p, u_1) + \mu y_m(t, p, u_2).$$

Control theory usually refers to this type of nonlinearity. When linear models are considered, time invariance is usually also assumed implicitly, i.e. the behavior of the model is assumed invariant by translation of the origin of time.

A model is *nonlinear with respect to its parameters (non-LP)* if its output does not satisfy the superposition principle with respect to its parameters, i.e. if

$$\exists \lambda, \mu \in \mathbb{R}, t \in \mathbb{R}^+ |y_m(t, \lambda p_1 + \mu p_2, u) \neq \lambda y_m(t, p_1, u) + \mu y_m(t, p_2, u).$$

In a *statistical context*, nonlinear models usually refer to this type of nonlinearity.

A model is *affine with respect to its parameters (AP)* if its output satisfies

$$y_m(t, p, u) = y_{m_1}(t, u) + y_{m_2}(t, p, u),$$

with $y_{m_2}(t, p, u)$ LP. AP models differ from LP models by the addition of a term that does not depend on the parameters, so that all methods available for the study of LP models extend to AP models without any difficulty. In what follows, we shall indulge in calling LP any LP or AP model.

Let us consider the consequences of nonlinearity on the structural identifiability of a model. If $M(.)$ is LP (or AP), the identity of the input/output behaviors results in a set of linear equations in $\hat{p}$

$$M(\hat{p}) = M(p^*) \iff A(t, u)\hat{p} = A(t, u)p^*.$$

Depending on the nature of the kernel of $A(t, u)$, this set of equations may have a unique solution in $\hat{p} = p^*$ or an uncountable set of solutions. Any s.l.i. LP model is therefore s.g.i. too, and a local study is sufficient to get global results.

On the other hand, if the model is non-LP four cases must be distinguished:

$$M(\hat{p}) = M(p^*) \implies \{\hat{p}\} = p^* \implies M(.) \text{ is s.g.i.}$$
$$M(\hat{p}) = M(p^*) \implies \{\hat{p}\} \text{ finite or countable} \implies M(.) \text{ is s.l.i.}$$
$$M(\hat{p}) = M(p^*) \implies \{\hat{p}\} \text{ uncountable} \implies M(.) \text{ is s.n.i.}$$
$$M(\hat{p}) = M(p^*) \text{ does not lead to any structural conclusion.}$$

The nonlinear dependency of the model output in the parameters therefore forbids reaching a global conclusion from a local study, and global tools will be needed. In what follows, only non-LP models will be considered, for they are the only ones that raise any difficulty. The techniques to be used will differ depending on whether the model studied is LI or not. Let us start with the LI case, which is easier.

3.4 Methods of test for LI models

Consider a model structure described by the time-invariant LI state-space equation

$$M(p) : \begin{cases} \dfrac{\mathrm{d}}{\mathrm{d}t}x = A(p)x + B(p)u,\, x(0) = x_0(p), \\[2mm] y_m(t,p) = C(p)x. \end{cases}$$

For some types of structures, such as mamillary or catenary compartmental models (often used in biology), general results exist that allow a conclusion to be drawn about their identifiability without any calculation (see, e.g., Anderson, 1982; Audoly and d'Angio, 1983; Cobelli, Lepschy and Romanin Jacur, 1979 a, b). In all other cases, it is necessary to perform a specific study. The most commonly used tools are the Laplace transform approach (Bellman and Åström, 1970) and the state-space similarity transformations approach (Berman and Schoenfeld, 1956; Glover and Willems, 1974; Walter and Lecourtier, 1981).

3.4.1 Laplace transform approach

Denote the Laplace variable by s. The Laplace transform of the previous equation can be written

$$Y_m(s,p) = H_1(s,p)x_0(p) + H_2(s,p)U(s),$$

with

$$H_1(s,p) = C(p)[sI - A(p)]^{-1} \text{ and } H_2(s,p) = C(p)[sI - A(p)]^{-1}B(p).$$

The models $M(\hat{p})$ and $M(p^*)$ will have the same input/output behavior for all times and inputs if and only if

$$Y_m(s,\hat{p}) - Y_m(s,p^*) \equiv 0 \quad \forall\, s, U(s).$$

This results in a set of equations binding $\hat{p}$ and p^*. By computing the number of solutions for $\hat{p}$ of these equations, one can conclude on the structural identifiability of $M(.)$.

Example Consider the model structure defined by

$$M(p) : \begin{cases} \dfrac{\mathrm{d}}{\mathrm{d}t}x_1 = -(p_1 + p_2)x_1 + p_3x_2 + u, & x_1(0) = 0, \\[2mm] \dfrac{\mathrm{d}}{\mathrm{d}t}x_2 = p_1x_1 - p_3x_2, & x_2(0) = 0, \\[2mm] y_m = x_2. \end{cases}$$

The corresponding transfer function is

$$H_2(s,p) = \frac{p_1}{s^2 + s(p_1 + p_2 + p_3) + p_2p_3}.$$

Since initial conditions are zero, the relation $M(\hat{p}) = M(p^*)$ translates into the equations

$$\begin{cases} \hat{p}_1 = p_1^*, \\ \hat{p}_1 + \hat{p}_2 + \hat{p}_3 = p_1^* + p_2^* + p_3^*, \\ \hat{p}_2\hat{p}_3 = p_2^*p_3^*. \end{cases}$$

This set of equations possesses two solutions for $\hat{p}$:

$$\hat{p}_1 = (p_1^*, p_2^*, p_3^*)^T \qquad \text{and} \qquad \hat{p}_2 = (p_1^*, p_3^*, p_2^*)^T.$$

The first parameter, which takes the same value in both solutions is s.g.i., the second and third parameters, which can each take two different values, are only s.l.i. From noise-free data, it will only be possible to uniquely determine the value of p_1, but for p_2 and p_3 one will obtain two possible values between which it will be impossible to choose without resorting to other measurements or prior knowledge than those assumed available during the structural identifiability study.

Remark 2

(i) Had we estimated the parameters of such a model (e.g. by minimizing a quadratic criterion) with any of the local methods available in commercial packages for data analysis, we would have found one of the two possible solutions and might have ignored the existence of the other one leading to exactly the same input/output behavior. Knowing either of the posterior feasible solutions, we are now able to generate the other, and thus to translate the ambiguity of the data at the model level.

(ii) This example illustrates the necessity of inserting the statement "for almost any value of p^*" in the definitions of structural identifiability and distinguishability. It is indeed clear from the expression of the transfer function H_2 that if p^* belongs to the plane of the parametric space defined by $p_1 = 0$, then the output of the "process" will be identically zero whatever the input, so that p_2 and p_3 will become non-identifiable. One will nevertheless say that they are s.l.i., because the plane $p_1 = 0$ can be considered as atypical.

(iii) The fact that this model is not s.g.i. implies that it will not be possible to estimate uniquely its state $x(t)$ from the sole knowledge of its input/output behavior. Depending on the model selected, one will indeed obtain two possible values for x_1. If one is interested in the evolution of a vector z that contains some not directly measured state variables, it is therefore advisable to check that the model is s.g.i., or at least that $M(\hat{p}) = M(p^*)$ implies $z_m(t, \hat{p}) \equiv z_m(t, p^*)$.

(iv) It is actually not necessary to assume the existence of a true value for the parameters; p^* may be viewed as the vector of the parameters of a model generating a satisfactory input/output behavior (which could have been obtained by a traditional estimation algorithm). The problem is then to know whether there exist other models with the same structure that have the same input/output behavior as this generating model.

(v) To avoid calculation errors, it is advisable to put each entry of the transfer matrices H_1 and H_2 into some canonical form, i.e. a form such that there is only one way of writing it. One obtains, for example, a canonical form by writing each entry of the transfer matrix as the ratio of two ordered polynomials in s, provided that these polynomials are divided by their GCD and that the coefficient of the denominator with lowest (or highest) degree in s is set equal to one.

3.4.2 State-space similarity transformation approach

Consider a model structure defined by

$$M(p^*) : \begin{cases} \dfrac{d}{dt} x^* = A(p^*)x^* + B(p^*)u, & x^*(0) = x_0(p^*), \\ y_m = C(p^*)x^*, \end{cases}$$

and set

$$\hat{x} = Tx^*,$$

where T is a constant invertible square matrix. Then

$$\begin{cases} \dfrac{d}{dt}\hat{x} = TA(p^*)T^{-1}\hat{x} + TB(p^*)u, \hat{x}(0) = Tx_0(p^*), \\ y_m = C(p^*)T^{-1}\hat{x}. \end{cases} \tag{3.1}$$

If

$$\begin{cases} A(\hat{p}) = TA(p^*)T^{-1}, \\ B(\hat{p}) = TB(p^*), \\ C(\hat{p}) = C(p^*)T^{-1}, \\ x_0(\hat{p}) = Tx_0(p^*), \end{cases} \tag{3.2}$$

then (3.1) describes the model $M(\hat{p})$ and

$$M(\hat{p}) = M(p^*).$$

From Kalman's algebraic equivalence theorem, the converse is also true, provided that $M(.)$ is structurally controlable and structurally observable. One can then test the structural identifiability of $M(.)$ by looking for all solutions of (3.2) for $(\hat{p}, T)$. If for almost any p^* the only solution is $(\hat{p}, T) = (p^*, I)$, then $M(.)$ is s.g.i. If for almost any p^* the set of solutions for $\hat{p}$ is finite or countable, then $M(.)$ is s.l.i.

Example Consider again the previous example. The model structure is defined by

$$A(p) = \begin{bmatrix} -(p_1 + p_2) & p_3 \\ p_1 & -p_3 \end{bmatrix}, B(p) = \begin{bmatrix} 1 \\ 0 \end{bmatrix}, C(p) = [0\ 1], x(0) = 0.$$

It corresponds to structurally controlable and observable models, so that the similarity transformation approach applies. Initial conditions bring no information. Let $T = [t_{ij}]$. The structure of the control and observation matrices implies

$$\begin{aligned} C(\hat{p})T &= C(p^*) &\implies t_{21} = 0, t_{22} = 1, \\ B(\hat{p}) &= TB(p^*) &\implies t_{11} = 1. \end{aligned}$$

The similarity transformation matrix can therefore be written

$$T(\alpha) = \begin{bmatrix} 1 & \alpha \\ 0 & 1 \end{bmatrix};$$

and the set of possible matrices A is given by

$$A(\hat{p}) = T(\alpha)A(p^*)T^{-1}(\alpha) = \begin{bmatrix} -(p_1^* + p_2^*) + \alpha p_1^* & \alpha(p_1^* + p_2^*) + p_3^* - \alpha^2 p_1^* - \alpha p_3^* \\ p_1^* & -\alpha p_1^* - p_3^* \end{bmatrix}.$$

For the structure of $A(\hat{p})$ to be correct, the sum of the entries of its second column must be zero, which implies

$$\alpha^2 p_1^* + \alpha(p_3^* - p_2^*) = 0.$$

This equation possesses two solutions for α, namely:

$$\alpha = 0 \Longrightarrow T = I, \hat{p} = p^*$$

and

$$\alpha = \frac{p_2^* - p_3^*}{p_1^*} \Longrightarrow T \neq I, \hat{p} = \begin{bmatrix} p_1^* \\ p_3^* \\ p_2^* \end{bmatrix}.$$

The same conclusion as before is therefore obtained. $M(.)$ is s.l.i., only p_1 is s.g.i., p_2 and p_3 may be exchanged without modifying the input/output behavior. From ideal data, it is impossible to uniquely estimate p^* or x^*, but we now know how to calculate all their possible values.

Remark 3

Even if the conclusion does not depend on the method used to obtain it, the calculations to be performed do. One may therefore naturally wonder which method is simpler to apply. This question unfortunately receives no universal answer. It is easy to generate examples for which solution is trivial with either approach and much more complex with the other. This situation extends to the non-LI case (Chappell, Godfrey and Vajda, 1990). It is therefore important to have varied tools at one's disposal in order to be able to choose the most appropriate one on a case-by-case basis.

3.5 Methods of test for non-LI models

3.5.1 Taylor series approach

Consider the model structure defined by

$$M(p) : \begin{cases} \dfrac{d}{dt}x(t) = f(x(t), u(t), t, p), \; x(0) = x_0(p), \\ y_m(t, p) = h(x(t), p), \end{cases}$$

where f and h are assumed to be infinitely continuously differentiable. Let

$$a_k(p) = \lim_{t \to 0+} \frac{d^k}{dt^k} y_m(t, p).$$

The fact that $M(\hat{p}) = M(p^*)$ implies

$$a_k(\hat{p}) = a_k(p^*), k = 0, 1, \ldots$$

A *sufficient* condition for $M(.)$ to be s.g.i. is therefore (Pohjanpalo, 1978) that

$$a_k(\hat{p}) = a_k(p^*), k = 0, 1, \ldots, k_{\max}, \Longrightarrow \hat{p} = p^*,$$

where $k_{\max}$ is any positive integer (small enough for the calculations to remain tractable).

Example Let us first see the results of this approach on the impulse response of an LI model initially at rest.

$$M(p) : \begin{cases} \dfrac{\mathrm{d}}{\mathrm{d}t}x &= A(p)x + B(p)u, \quad x(0) = 0, \\[2mm] y_m(t, p) &= C(p)x. \end{cases}$$

The matrix of the impulse responses of this model is given by

$$Y_m(t, p) = C(p) \, e^{[A(p)t]} B(p)$$

and

$$\lim_{t \longrightarrow 0+} \frac{\mathrm{d}^k}{\mathrm{d}t^k} Y_m(t, p) = C(p)A^k(p)B(p),$$

so that this approach amounts to testing identifiability from the identity of the Markov parameters (Fisher, 1966; Grewal and Glover, 1976)

$$C(\hat{p})A^k(\hat{p})B(\hat{p}) = C(p^*)A^k(p^*)B(p^*), k = 0, 1, \ldots, k_{max},$$

a method that usually turns out to require more calculations than the Laplace transform or similarity transformation approaches.

Example Consider now the non-LI model structure defined by

$$M(p) : \begin{cases} \dfrac{\mathrm{d}}{\mathrm{d}t}x &= \begin{bmatrix} -p_1 x_1 - p_2(1 - p_3 x_2)x_1 \\[1mm] p_2(1 - p_3 x_2)x_1 - p_4 x_2 \end{bmatrix}, \quad x(0) = \begin{bmatrix} 1 \\ 0 \end{bmatrix}, \\[4mm] y_m(t, p) &= x_1. \end{cases}$$

The successive derivatives of the model response evaluated at $t = 0^+$ are given by

$$\begin{aligned} a_0(p) &= 1 \\ a_1(p) &= -(p_1 + p_2) \\ a_2(p) &= (p_1 + p_2)^2 + p_2^2 p_3 \\ a_3(p) &= -p_2^3 p_3^2 - 4p_2^2 p_3(p_1 + p_2) - p_2^2 p_3 p_4 - (p_1 + p_2)^3 \end{aligned}$$

$$\cdots$$

One can easily show that

$$a_k(\hat{p}) = a_k(p^*), \qquad k = 1, 2, \ldots, 5 \Longrightarrow \hat{p} = p^*.$$

$M(.)$ is therefore s.g.i.

Remark 4

If p_3 is set to zero in the previous model structure, this structure becomes LI and s.n.i., which illustrates the fact, frequently recorded, that LI models tend to be less identifiable than non-LI models. For a long period, the only examples of non-LI and non-s.g.i. models that could be found in the literature were trivial (for example models containing parts that were not connected to the outputs). This reached the point where some doubted whether it would be of any practical interest to develop methods for testing non-LI models for structural identifiability. Hence the importance of non-trivial examples to caution experimentalists. Such examples will be provided in the following sections.

3.5.2 Generating series approach

This approach (Lecourtier, Lamnabhi-Lagarrigue and Walter, 1987; Walter, 1982) assumes the model to be written as

$$
M(p): \begin{cases}
\dfrac{\mathrm{d}}{\mathrm{d}t}x(t) = f_0(x(t), p) + \displaystyle\sum_{i=1}^{m} u_i(t) f_i(x(t), p)), & x(0) = x_0(p), \\
y_m(t, p) = h(x(t), p),
\end{cases}
$$

The state x evolves on a manifold $\mathbb{V}$ of $\mathbb{R}^n$, and $f_i (i = 0, \ldots, m)$ and h are assumed to be analytical on $\mathbb{V}$. The output can therefore be expanded in series with respect to time and inputs (Fliess, 1981)

$$
y_m(t, p) = h(x(t), p)|_0 + \sum_{k \geq 0} \sum_{j_0, \ldots, j_k = 0}^{m} L_{f_{j_0}} \cdots
$$
$$
L_{f_{j_{k-1}}} L_{f_{j_k}} h(x(t), p)|_0 \int_0^t \mathrm{d}\xi_{j_k}(\tau) \int_0^\tau \mathrm{d}\xi_{j_{k-1}} \cdots \mathrm{d}\xi_{j_0}.
$$

In this expresssion, $\mathrm{d}\xi_0 = \mathrm{d}t$ and $\mathrm{d}\xi_i(\tau) = u_i(\tau)\mathrm{d}\tau (i = 1, \ldots, m)$; $|_0$ indicates evaluation at the initial state $x(0)$ and $L_f h$ is the Lie derivative of h along the vector field f, given by

$$
L_f h(x(t), p) = \sum_{j=1}^{n} f_j(x(t), p) \frac{\partial}{\partial x_j} h(x(t), p),
$$

with f_j the jth component of f. Manipulating iterated integrals is difficult, and facilitated by using an operational calculus involving formal series in non-commutative variables. The generating series g_x associated with x is obtained with the help of the transformations

$$
x(t) \longrightarrow g_x,
$$
$$
x(0) \longrightarrow x(0),
$$
$$
\int_0^t x(\tau)\mathrm{d}\tau \longrightarrow v_0 g_x,
$$
$$
\int_0^t u_i(\tau)\mathrm{d}\tau \longrightarrow v_i,
$$
$$
\int_0^t u_i(\tau)x(\tau)\mathrm{d}\tau \longrightarrow v_i g_x,
$$
$$
\int_0^t \mathrm{d}\xi_{j_k}(\tau)) \int_0^\tau \mathrm{d}\xi_{j_{k-1}} \cdots \mathrm{d}\xi_{j_0} \longrightarrow v_{j_k} \cdots v_{j_0}.
$$

The *letters* $v_i (i = 0, \ldots, m)$ constitute *words* by (non-commutative) concatenation. The generating series g_y associated with the output y_m satisfies

$$g_y = h(x(t), p)|_0 + \sum_{k \geq 0} \sum_{j_0, \ldots, j_k = 0}^{m} L_{f_{j_0}} \ldots L_{f_{j_k}} h(x(t), p)|_0 \, v_{j_k} \ldots v_{j_0}.$$

$L_{f_{j_0}} \ldots L_{f_{j_k}} h(x(t), p)|_0$ is the coefficient of the generating series g_y associated with the word $v_{j_k} \ldots v_{j_0}$. It only depends on p. Let $s(p)$ be the vector of all coefficients of g_y. $M(\hat{p}) = M(p^*)$ translates into $s(\hat{p}) = s(p^*)$. One can therefore test the structural identifiability of $M(.)$ by calculating the number of solutions for $\hat{p}$ of the set of equations $s(\hat{p}) = s(p^*)$.

Example Let us first apply this approach to an LI model, which can be rewritten as

$$M(p) : \begin{cases} \dfrac{d}{dt} x(t) &= A(p)x(t) + \displaystyle\sum_{i=1}^{m} u_i(t) b_i(p), \qquad x(0) = x_0(p), \\ y_m(t, p) &= C(p) x(t). \end{cases}$$

Integrating the state equation once with respect to time, one obtains

$$x(t) = x(0) + A(p) \int_0^t x(\tau) d\tau + \sum_{i=1}^{m} b_i(p) \int_0^t u_i(\tau) d\tau,$$

$$\implies g_x = x(0) + A(p) v_0 g_x + \sum_{i=1}^{m} b_i(p) v_i.$$

The observation equation implies

$$g_y = C(p) g_x.$$

The output generating series is therefore given by

$$g_y = C(p)[I - A(p)v_0]^{-1} x_0(p) + C(p)[I - A(p)v_0]^{-1} \sum_{i=1}^{m} b_i(p) v_i,$$

identical, up to a change of variable, to the expression obtained with the Laplace transform approach:

$$Y_m(s, p) = C(p)[sI - A(p)]^{-1} x_0(p) + C(p)[sI - A(p)]^{-1} B(p) U(s).$$

For LI models, these two approaches are therefore equivalent.

In order to compute elements of $s(p)$, one may apply the definition and calculate terms of the type $L_{f_{j_0}} \ldots L_{f_{j_k}} h(x(t), p)|_0$. This is the procedure followed in the next example.

Example Consider the model structure defined by

$$M(p) : \begin{cases} \dfrac{d}{dt} x &= \begin{bmatrix} -\left(p_1 + \dfrac{p_2}{p_3 + x_1}\right) x_1 + p_4 x_2 \\ p_1 x_1 - p_4 x_2 \end{bmatrix} + \begin{bmatrix} 1 \\ 0 \end{bmatrix} u, \qquad x(0) = \mathbf{0}, \\ y_m(t, p) &= x_1. \end{cases}$$

It corresponds to

$$
\boldsymbol{f}_0 = \begin{bmatrix} -\left(p_1 + \dfrac{p_2}{p_3 + x_1}\right)x_1 + p_4 x_2 \\[2mm] p_1 x_1 - p_4 x_2 \end{bmatrix}
$$

$$
\Longrightarrow \quad L_{\boldsymbol{f}_0} = \left[-\left(p_1 + \frac{p_2}{p_3 + x_1}\right)x_1 + p_4 x_2 \right]\frac{\partial}{\partial x_1} + [p_1 x_1 - p_4 x_2]\frac{\partial}{\partial x_2},
$$

$$
\boldsymbol{f}_1 = \begin{bmatrix} 1 \\ 0 \end{bmatrix}
$$

$$
\Longrightarrow \quad L_{\boldsymbol{f}_1} = \frac{\partial}{\partial x_1},
$$

$$
\boldsymbol{h}(\boldsymbol{x}, \boldsymbol{p}) = x_1.
$$

Therefore

$$
\left.\begin{aligned}
L_{f_1} L_{f_0} h|_0 &= -\left(p_1 + \frac{p_2}{p_3}\right) \\[2mm]
L_{f_1} L_{f_1} L_{f_0} h|_0 &= 2\frac{p_2}{p_3^2} \\[2mm]
L_{f_1} L_{f_1} L_{f_1} L_{f_0} h|_0 &= -6\frac{p_2}{p_3^3} \\[2mm]
L_{f_1} L_{f_0} L_{f_0} h|_0 &= p_1 p_4 + \left(p_1 + \frac{p_2}{p_3}\right)^2
\end{aligned}\right\} \quad s(\hat{\boldsymbol{p}}) = s(\boldsymbol{p}^*) \Rightarrow \hat{\boldsymbol{p}} = \boldsymbol{p}^* \Rightarrow M(.) \text{ is s.g.i.}
$$

Note that the calculations would have been much more complex with the Taylor series approach.

This policy of generating the coefficients associated with words that are chosen more or less at random may multiply useless calculations, because many coefficients may turn out to be zero and thus bring no information on the parameters. It is therefore interesting to have at one's disposal systematic techniques for generating words associated with non-zero coefficients. Consider the case of bilinear models described by

$$
M(\boldsymbol{p}) : \begin{cases} \dfrac{\mathrm{d}}{\mathrm{d}t}\boldsymbol{x} &= A_0(\boldsymbol{p})\boldsymbol{x} + \displaystyle\sum_{i=1}^{m} u_i(A_i(\boldsymbol{p})\boldsymbol{x} + b_i(\boldsymbol{p})), \boldsymbol{x}(0) = \boldsymbol{x}_0(\boldsymbol{p}), \\[3mm] \boldsymbol{y}_m(t, \boldsymbol{p}) &= C(\boldsymbol{p})\boldsymbol{x}(t). \end{cases}
$$

Integrating the state equation once with respect to t, one gets

$$
\boldsymbol{x}(t) - \boldsymbol{x}_0(\boldsymbol{p}) = A_0(\boldsymbol{p})\int_0^t \boldsymbol{x}(\tau)\mathrm{d}\tau + \sum_{i=1}^{m}\int_0^t u_i(\tau)(A_i(\boldsymbol{p})\boldsymbol{x}(\tau) + b_i(\boldsymbol{p}))\mathrm{d}\tau
$$

$$
\Longrightarrow \boldsymbol{g}_x = \boldsymbol{x}_0(\boldsymbol{p}) + v_0 A_0(\boldsymbol{p})\boldsymbol{g}_x + \sum_{i=1}^{m} v_i(A_i(\boldsymbol{p})\boldsymbol{g}_x + b_i(\boldsymbol{p}))
$$

$$
\Longrightarrow \boldsymbol{g}_x = (I - v_0 A_0(\boldsymbol{p}))^{-1}\boldsymbol{x}_0(\boldsymbol{p}) + (I - v_0 A_0(\boldsymbol{p}))^{-1}\sum_{i=1}^{m} v_i[A_i(\boldsymbol{p})\boldsymbol{g}_x + b_i(\boldsymbol{p})].
$$

Let ${}^{k}\boldsymbol{g}_{\boldsymbol{x}}$ be the sum of all terms of $\boldsymbol{g}_{\boldsymbol{x}}$ that contain k occurrences of the letters $v_i(i = 1,\ldots,m)$ and an arbitrary number of occurrences of v_0. Then

$$
{}^{0}\boldsymbol{g}_{\boldsymbol{x}} = (\boldsymbol{I} - v_0 \boldsymbol{A}_0(\boldsymbol{p}))^{-1} \boldsymbol{x}_0(\boldsymbol{p}),
$$

$$
{}^{1}\boldsymbol{g}_{\boldsymbol{x}} = (\boldsymbol{I} - v_0 \boldsymbol{A}_0(\boldsymbol{p}))^{-1} \sum_{i=1}^{m} v_i[\boldsymbol{A}_i(\boldsymbol{p})({}^{0}\boldsymbol{g}_{\boldsymbol{x}}) + \boldsymbol{b}_i(\boldsymbol{p})],
$$

$$
{}^{k}\boldsymbol{g}_{\boldsymbol{x}} = (\boldsymbol{I} - v_0 \boldsymbol{A}_0(\boldsymbol{p}))^{-1} \sum_{i=1}^{m} v_i \boldsymbol{A}_i(\boldsymbol{p})({}^{k-1}\boldsymbol{g}_{\boldsymbol{x}})) \text{ for } k \geq 2.
$$

One thus obtains a recurrence equation allowing the calculation of all nonzero terms of $\boldsymbol{g}_{\boldsymbol{x}}$. Note that the term $(\boldsymbol{I} - v_0 \boldsymbol{A}_0(\boldsymbol{p}))^{-1}$ can be computed once and for all.

Example   Consider the bilinear model structure defined by

$$
M(\boldsymbol{p}) : \begin{cases} \dfrac{d}{dt}\boldsymbol{x} = \begin{bmatrix} -p_1 x_1 + p_2 u \\ -p_3 x_2 + p_4 u \\ -(p_1 + p_3)x_3 + (p_4 x_1 + p_2 x_2)u \end{bmatrix} \\[2em] \boldsymbol{x}(0) = \boldsymbol{0} \\[0.5em] y_m(t,\boldsymbol{p}) = x_3. \end{cases}
$$

With the previous notation, it corresponds to

$$
\boldsymbol{b}_1 = \begin{bmatrix} p_2 \\ p_4 \\ 0 \end{bmatrix}, \boldsymbol{A}_0 = \begin{bmatrix} -p_1 & 0 & 0 \\ 0 & -p_3 & 0 \\ 0 & 0 & -(p_1 + p_3) \end{bmatrix}, \boldsymbol{A}_1 = \begin{bmatrix} 0 & 0 & 0 \\ 0 & 0 & 0 \\ p_4 & p_2 & 0 \end{bmatrix}
$$

$$
\implies (\boldsymbol{I} - v_0 \boldsymbol{A}_0)^{-1} = \begin{bmatrix} (1 + p_1 v_0)^{-1} & 0 & 0 \\ 0 & (1 + p_3 v_0)^{-1} & 0 \\ 0 & 0 & (1 + (p_1 + p_3)v_0)^{-1} \end{bmatrix}
$$

Since the initial conditions are zero, ${}^{0}\boldsymbol{g}_{\boldsymbol{x}} = \boldsymbol{0}$. The recurrence then implies

$$
{}^{1}\boldsymbol{g}_{\boldsymbol{x}} = \begin{bmatrix} (1 + p_1 v_0)^{-1} p_2 v_1 \\ (1 + p_3 v_0)^{-1} p_4 v_1 \\ 0 \end{bmatrix},
$$

$$
{}^{2}\boldsymbol{g}_{\boldsymbol{x}} = \begin{bmatrix} 0 \\ 0 \\ p_2 p_4 (1 + (p_1 + p_3)v_0)^{-1}[v_1(1 + p_1 v_0)^{-1}v_1 + v_1(1 + p_3 v_0)^{-1}v_1] \end{bmatrix},
$$

$$
{}^{k}\boldsymbol{g}_{\boldsymbol{x}} = \boldsymbol{0} \quad \forall k > 2.
$$

The generating series $\boldsymbol{g}_y$ is obtained as the last component of $\boldsymbol{g}_{\boldsymbol{x}}$. Since p_2 and p_4 only appear by their product, each of them is s.n.i.

3.5.3 Local state isomorphism approach

Consider a model structure defined by

$$
M(p) : \begin{cases}
\dfrac{\mathrm{d}}{\mathrm{d}t}x(t) &= f(x(t), p) + u(t)g(x(t), p), \\[2mm]
x(0) &= x_0(p), \\[2mm]
y_m(t, p) &= h(x(t), p),
\end{cases}
$$

where f, g and h are real analytic over the manifold $\mathbb{V}$ of $\mathbb{R}^n$ on which x evolves, where the input u belongs to the set of measurable bounded functions and where $M(p)$ is locally reduced at $x_0(p)$ for almost any p (which corresponds to a notion of structural observability and structural controlability (Hermann and Krener, 1977; Sussmann, 1977)). Let x^* be the state of $M(p^*)$ and $\hat{x}$ that of $M(\hat{p})$. The local state isomorphism theorem (Hermann and Krener, 1977; Sussmann, 1977; Tunali and Tarn, 1987) implies that $M(p^*)$ and $M(\hat{p})$ will have the same input/output behavior for all u up to some time $t_1 > 0$ if and only if there exists a local state isomorphism

$$
\phi : v(x_0^*) \longrightarrow \mathbb{R}^n
$$
$$
x^* \longmapsto \hat{x} = \phi(x^*)
$$

such that *for any* x^* belonging to the neighborhood $v(x_0^*)$ the following conditions be satisfied:

(i) ϕ is a diffeomorphism

$$
\operatorname{rank} \frac{\partial}{\partial x^T}\phi(x)|_{x=x^*} = n \, ;
$$

(ii) the initial states correspond

$$
\phi(x_0^*) = \hat{x}_0 \, ;
$$

(iii) the drift terms correspond

$$
\hat{f}(\hat{x}) = \hat{f}(\phi(x^*)) = \frac{\partial}{\partial x^T}\phi(x)|_{x=x^*} f^*(x^*) \, ;
$$

(iv) the control terms correspond

$$
\hat{g}(\hat{x}) = \hat{g}(\phi(x^*)) = \frac{\partial}{\partial x^T}\phi(x)|_{x=x^*} g^*(x^*) \, ;
$$

(v) the observations correspond

$$
\hat{h}(\hat{x}) = \hat{h}(\phi(x^*)) = h^*(x^*).
$$

In order to apply this theorem to identifiability testing, it suffices, after having checked that $M(p)$ is locally reduced at $x_0(p)$ with the help of techniques described by Hermann and Krener (1977), to use conditions (i)–(v) to find the set of all solutions for $\hat{p}$ and ϕ (Vajda and Rabitz, 1989; Vajda, Godfrey and Rabitz, 1989). If for almost any p^* the only solution is the trivial solution $\hat{p} = p^*$, $\phi(x^*) = x^*$, then M is s.g.i. If the set of all solutions for $\hat{p}$ is finite or countable, then M is s.l.i.

Before treating non-LI examples, let us see what becomes of this approach in the LI case. The local state isomorphism becomes a similarity transformation $\phi(x^*) = Tx^*$ and conditions (i) to (v) respectively become

$$\text{rank } T = n,$$
$$Tx_0^* = \hat{x}_0,$$
$$A(\hat{p})Tx^* = TA(p^*)x^*,$$
$$B(\hat{p}) = TB(p^*),$$
$$C(\hat{p})Tx^* = C(p^*)x^*.$$

They correspond to the conditions obtained with the state-space similarity transformation approach.

Example   Consider the model structure with two state variables defined by

$$f(x, p) = \begin{bmatrix} p_1 x_1^2 + p_2 x_1 x_2 \\ p_3 x_1^2 + p_4 x_1 x_2 \end{bmatrix}, g(x, p) = \begin{bmatrix} 1 \\ 0 \end{bmatrix}, h(x, p) = x_1 \text{ and } x_0 = 0,$$

which we shall admit to be locally reduced. (For a proof, see Vajda, Godfrey and Rabitz (1989)). Condition (v) implies

$$\hat{x}_1 = \phi_1(x^*) = x_1^* \implies \frac{\partial}{\partial x_1}\phi_1(x)_{|x=x^*} = 1, \frac{\partial}{\partial x_2}\phi_1(x)_{|x=x^*} = 0.$$

Condition (iv) results in

$$\begin{bmatrix} 1 \\ 0 \end{bmatrix} = \frac{\partial}{\partial x^T}\phi(x)_{|x=x^*} \begin{bmatrix} 1 \\ 0 \end{bmatrix} = \begin{bmatrix} \frac{\partial}{\partial x_1}\phi_1(x)_{|x=x^*} \\ \frac{\partial}{\partial x_1}\phi_2(x)_{|x=x^*} \end{bmatrix} \implies \frac{\partial}{\partial x_1}\phi_2(x)_{|x=x^*} = 0.$$

Taking the previous results into account, one can rewrite condition (iii) as

$$\begin{bmatrix} \hat{p}_1\hat{x}_1^2 + \hat{p}_2\hat{x}_1\hat{x}_2 \\ \hat{p}_3\hat{x}_1^2 + \hat{p}_4\hat{x}_1\hat{x}_2 \end{bmatrix} = \begin{bmatrix} 1 & 0 \\ 0 & \frac{\partial}{\partial x_2}\phi_2(x)_{|x=x^*} \end{bmatrix} \begin{bmatrix} p_1^* x_1^{*2} + p_2^* x_1^* x_2^* \\ p_3^* x_1^{*2} + p_4^* x_1^* x_2^* \end{bmatrix}.$$

Since $\hat{x}_1 = x_1^*$ and $\hat{x}_2 = \phi_2(x^*)$, the first row implies

$$\hat{p}_1 x_1^{*2} + \hat{p}_2 x_1^* \phi_2(x^*) = p_1^* x_1^{*2} + p_2^* x_1^* x_2^*, \quad \forall x^* \in v(x_0^*).$$

Since $\frac{\partial}{\partial x_1}\phi_2(x)_{|x=x^*} = 0$, $\phi_2(x^*)$ is not affine in x_1^*. The identity of the terms in x_1^{*2} therefore implies

$$\hat{p}_1 = p_1^*,$$

and thus

$$\phi_2(x^*) = \frac{p_2^*}{\hat{p}_2}x_2^* \implies \frac{\partial}{\partial x_2}\phi_2(x)_{|x=x^*} = \frac{p_2^*}{\hat{p}_2}.$$

The second row can then be rewritten as

$$\hat{p}_3 x_1^{*2} + \hat{p}_4 x_1^* \frac{p_2^*}{\hat{p}_2} x_2^* = \frac{p_2^*}{\hat{p}_2} p_3^* x_1^{*2} + \frac{p_2^*}{\hat{p}_2} p_4^* x_1^* x_2^*, \quad \forall x^* \in v(x_0^*),$$

which implies

$$\hat{p}_3 = \frac{p_2^*}{\hat{p}_2} p_3^* \iff \hat{p}_2 \hat{p}_3 = p_2^* p_3^*$$

and

$$\hat{p}_4 \frac{p_2^*}{\hat{p}_2} = \frac{p_2^*}{\hat{p}_2} p_4^* \iff \hat{p}_4 = p_4^*.$$

Condition (ii) translates into

$$\phi(0) = 0$$

and condition (i) can now be written

$$\operatorname{rank} \begin{bmatrix} 1 & 0 \\ 0 & \dfrac{p_2^*}{\hat{p}_2} \end{bmatrix} = 2 \iff p_2^* \neq 0,$$

which is structurally true. The parameters p_1 and p_4 are therefore s.g.i. while p_2 and p_3 are s.n.i. (only their product is s.g.i.). $M(.)$ is therefore s.n.i.

Example Consider the model structure with two state variables defined by

$$f(x, p) = \begin{bmatrix} p_1 x_1 x_2 \\ p_2 x_1 x_2 + p_3 x_2^2 \end{bmatrix}, \quad g(x, p) = \begin{bmatrix} 1 \\ 0 \end{bmatrix}, \quad h(x, p) = x_2 \text{ and } x_0 = \begin{bmatrix} p_4 \\ y(0) \end{bmatrix},$$

with $y(0) \neq 0$. The same procedure can be used (Vajda, Godfrey and Rabitz, 1989) to show that this model is s.l.i. and that there are two values of the parameter vector that lead to the same input-output behavior, namely $\hat{p} = p^*$ and

$$\begin{bmatrix} \hat{p}_1 \\ \hat{p}_2 \\ \hat{p}_3 \\ \hat{p}_4 \end{bmatrix} = \begin{bmatrix} p_3^* \\ p_2^* \\ p_1^* \\ p_4^* + \dfrac{p_3^* - p_1^*}{p_2^*} y(0) \end{bmatrix}$$

Only p_2 is therefore s.g.i. Note that the model becomes globally identifiable on the atypical manifold $p_3^* = p_1^*$, but this is a non-structural result.

The two previous examples are particular cases of polynomial models with linear observations, i.e. of models for which the components of f and g can be written as polynomials in x parametrized by p and for which $h(x(t, p), p) = C(p)x(t, p)$. For such models, one can directly write ϕ under the form of a linear transformation $\hat{x} = Tx^*$ (Chappell,

Godfrey and Vajda, 1990). Conditions (i)–(v) then become

(i') $\quad \det T \neq 0,$

(ii') $\quad T x_0^* = \hat{x}_0,$

(iii') $\quad \hat{f}(T x^*) = T f^*(x^*),$

(iv') $\quad \hat{g}(T x^*) = T g^*(x^*),$

(v') $\quad C(\hat{p}) T x^* = C(p^*) x^*,$

which notably simplify calculations. Vajda *et al.* (1989) applied this approach to a non-LI model of methane pyrolisis.

3.6 Contributions of computer algebra and elimination theory

Structural identifiability and distinguishability testing often involves conceptually simple but practically very complex algebraic manipulations, as could already be suspected from the simple illustrative examples treated so far. Examples of more realistic size require computations that challenge the patience of the most dedicated researcher. Hence the interest in executing these manipulations with a computer with the help of a computer algebra language, such as REDUCE, MACSYMA, MAPLE or AXIOM (Chappell, Godfrey and Vajda, 1990; Launay, 1989; Lecourtier and Raksanyi, 1987; Walter and Lecourtier, 1982). The first task to be performed is to deduce from the considered model structures, a set of equations binding $\hat{p}$ and p^* and expressing the identity of the input/output behaviors. For instance, to obtain the transfer matrix associated with an LI state-space model in canonical form, it suffices in REDUCE to type

```
ON GCD;
HC := C * (1/(s * IM − A)) * B;
```

after having defined the matrices A, B and C of the model and the identity matrix IM (In practice, it is possible that this seemingly simple operation already results in computations that are too complex to be executed and that a more sophisticated technique avoiding the inversion of $sI − A$ needs to be used.). Once the equations expressing the identity of the input/output behaviors have been obtained, *all* their solutions for $\hat{p}$ must be computed. This usually amounts to solving a set of polynomial equations in $\hat{p}$ parametrized by p^*, which may be written as

$$\begin{cases} P_1(\hat{p}, p^*) = 0, \\ \vdots \\ P_{n_p}(\hat{p}, p^*) = 0. \end{cases}$$

Tools of elimination theory make it possible to transform this set of equations into one that possesses the same solutions but is simpler to solve. A possible algorithm is as follows:

(i) In each polynomial, order monomials by decreasing degree in $\hat{p}$ (using, e.g., lexico-graphical ordering). Each polynomial equation can then be written as

$$P_i(\hat{p}) = M_i(\hat{p}) + R_i(\hat{p}) = 0 \qquad (i = 1, \ldots, n_p),$$

where the dependency in p^* has been omitted to simplify notation and where M_i is the monomial with the highest degree and R_i is the *reductum*.

(ii) Alternate reduction and construction phases.

Reduction phase

The polynomial P_j is said to be reduced with respect to P_i if $M_i(\hat{p})$ is replaced in it by $-R_i(\hat{p})$ whenever possible. This does not change the set of all solutions but decreases the degree of P_j (which, by the way, does not imply that the resulting expression looks simpler, etc.). Three cases must be distinguished. If $P_j(\hat{p}, p^*) \equiv 0$, this polynomial is redundant and can therefore be discarded from the set of equations to be considered. If $P_j(\hat{p}, p^*) = c(p^*) \neq 0$, the procedure can be terminated, for the set of equations has no solution for $\hat{p}$. Finally, if $P_j(\hat{p}, p^*)$ still contains components of $\hat{p}$, it must be kept.

Construction phase

When no further reduction is possible, a new polynomial equation is constructed according to the rule

$$P_{n_p+1}(\hat{p}, p^*) = k_i(\hat{p})P_j(\hat{p}, p^*) + k_j(\hat{p})P_i(\hat{p}, p^*) = 0.$$

Any solution $\hat{p}$ of the original set of equations is also a solution of this equation, so that appending this new equation to the set does not modify the solution set. Many policies can be considered for the choice of the polynomials i and j to be used as well as for the computation of $k_i(\hat{p})$ and $k_j(\hat{p})$. Buchberger (1970) suggests, for example, choosing the two polynomials with the lowest lexicographical degree that have not yet been selected together during a construction phase and to set

$$k_i(\hat{p}) = \frac{M_i(\hat{p})}{\text{GCD}\,(M_i(\hat{p}), M_j(\hat{p}))} \quad \text{and} \quad k_j(\hat{p}) = -\frac{M_j(\hat{p})}{\text{GCD}\,(M_i(\hat{p}), M_j(\hat{p}))},$$

which makes it possible to zero the coefficient of the monomial with the highest degree in P_{n_p+1}.

A reduction of P_{n_p+1} with respect to all other polynomials P_j is then performed. If P_{n_p+1} does not reduce to zero, it is then appended to the set of equations to be considered by incrementing n_p by one. All polynomials are then reduced with one another before switching back to a construction phase. The procedure is stopped when all polynomials that can be constructed reduce to zero, which can be shown to take place after a finite number of iterations. The system of equations thus obtained is much simpler to solve than the original one, since it is usually in triangular form, with the indeterminates appearing one by one.

Chemical engineering example (cont'd) The following set of equations has been obtained by the Laplace transform approach described in section 3.4.1. It expresses, after a change of variables aimed at making it polynomial (Walter, Piet-Lahanier and Happel, 1986), the identity of the input/output behaviors of the two model structures considered in section 3.1.

$$\begin{cases} \hat{p}_4 - \hat{p}_2 = p_4^*, \\ \hat{p}_1 + \hat{p}_3 + \hat{p}_4 = p_1^* + p_2^* + p_3^* + p_4^*, \\ \hat{p}_1\hat{p}_3 + \hat{p}_1\hat{p}_4 + \hat{p}_3\hat{p}_4 = p_1^*p_3^* + p_1^*p_4^* + p_2^*p_3^* + p_3^*p_4^*, \\ \hat{p}_1\hat{p}_3\hat{p}_4 = p_1^*p_3^*p_4^*. \end{cases}$$

By the reduction-construction procedure that has just been described, it can be put in the following triangular form with respect to $\hat{p}$:

$$\begin{cases} \hat{p}_2 - \hat{p}_4 + p_4^* = 0, \\ \hat{p}_1 + \hat{p}_3 + \hat{p}_4 - p_1^* - p_2^* - p_3^* - p_4^* = 0, \\ \hat{p}_3^2 + \hat{p}_3\,[\hat{p}_4 - p_1^* - p_2^* - p_3^* - p_4^*] + \hat{p}_4^2 - \hat{p}_4\,[p_1^* + p_2^* + p_3^* + p_4^*] + p_1^*p_3^* + p_1^*p_4^* \\ \quad + p_2^*p_3^* + p_3^*p_4^* = 0, \\ -\hat{p}_4^3 + \hat{p}_4^2\,[p_1^* + p_2^* + p_3^* + p_4^*] - \hat{p}_4\,[p_1^*p_3^* + p_1^*p_4^* + p_2^*p_3^* + p_3^*p_4^*] + p_1^*p_3^*p_4^* = 0. \end{cases}$$

The last of these two equations possesses three real solutions for $\hat{p}_4$. For each of them, the penultimate equation possesses two real solutions for $\hat{p}_3$. For each of the six possible combinations for $\hat{p}_3$ and $\hat{p}_4$ thus obtained, the first two equations possess a unique solution for $\hat{p}_1$ and $\hat{p}_2$, so that the system structurally has six solutions for $\hat{p}$.

Similarly, if one exchanges the roles of $\hat{p}$ and p^* one obtains the following triangular form with respect to p^*:

$$\begin{cases} p_4^* + \hat{p}_2 - \hat{p}_4 = 0, \\ p_1^*\,[\hat{p}_2 - \hat{p}_4] + p_3^{*2} - p_3^*\,[\hat{p}_1 + \hat{p}_3 + \hat{p}_4] + \hat{p}_1\hat{p}_3 + \hat{p}_1\hat{p}_4 + \hat{p}_3\hat{p}_4 = 0, \\ p_2^*\,[\hat{p}_2 - \hat{p}_4] - p_3^{*2} + p_3^*\,[\hat{p}_1 + \hat{p}_2 + \hat{p}_3] - \hat{p}_1\hat{p}_2 - \hat{p}_1\hat{p}_3 - \hat{p}_2^2 - \hat{p}_2\hat{p}_3 + \hat{p}_2\hat{p}_4 = 0, \\ p_3^{*3} - p_3^{*2}\,[\hat{p}_1 + \hat{p}_3 + \hat{p}_4] + p_3^*\,[\hat{p}_1\hat{p}_3 + \hat{p}_1\hat{p}_4 + \hat{p}_3\hat{p}_4] - \hat{p}_1\hat{p}_3\hat{p}_4 = 0. \end{cases}$$

The last equation of this set has three solutions for p_3^*. To each of them corresponds a unique solution for p_1^*, p_2^* and p_4^* of the three preceding equations. This set of equations therefore structurally has three solutions for p^*.

The two model structures considered are therefore not distinguishable. If reality is described by $M(p^*)$, it will be structurally possible to find six models with the other structure $\hat{M}(.)$ and the same external behavior. Conversely, if reality is described by $\hat{M}(\hat{p})$, it will be structurally possible to find three models with structure $M(.)$ and the same behavior. We deduce also from these results that neither $\hat{M}(.)$ nor $M(.)$ are s.g.i. They are only s.l.i. with respectively six and three different models having the same external behavior. There is, therefore, no hope of selecting the best model structure and estimating its parameters uniquely from the experiments considered. However, we can now easily generate the set of all models with either structure having the same behavior as any satisfactory generating model with either structure (Walter, Piet-Lahanier and Happel, 1986).

So far, we have used elimination theory to simplify the solution of sets of algebraic equations. Differential algebra (see, e.g., Fliess (1989)), in which differentiation is added to the classical axioms of algebra, allows one to extend this approach to eliminating state variables in order to obtain input/output differential equations (thus only involving known variables and the parameters to be estimated) from which identifiability can be studied (Ollivier, 1990), as illustrated by the following example.

Example Consider the bilinear structure defined by

$$
M(\boldsymbol{p}) : \begin{cases} \dfrac{d}{dt}\boldsymbol{x} = \begin{bmatrix} -(p_1 + p_2)x_1 + (p_5 u + p_3)x_2 \\ p_2 x_1 - (p_3 + p_4)x_2 \end{bmatrix} \\[4mm] \boldsymbol{x}(0) = \begin{bmatrix} 1 \\ 0 \end{bmatrix} \\[4mm] y_m(t, \boldsymbol{p}) = x_1. \end{cases}
$$

We wish to eliminate the state in order to obtain an input/output differential equation. Given the observation equation, it suffices to eliminate x_2. Differentiating the equation satisfied by $\dfrac{dx_1}{dt}$, with respect to time, we get

$$
\frac{d^2 y_m}{dt^2} + (p_1 + p_2)\frac{dy_m}{dt} - (p_5 u + p_3)\frac{dx_2}{dt} - p_5 x_2 \frac{du}{dt} = 0.
$$

Replacing $\dfrac{dx_2}{dt}$ by its value as given by the second component of the state equation, we obtain

$$
\frac{d^2 y_m}{dt^2} + (p_1 + p_2)\frac{dy_m}{dt} - (p_5 u + p_3)p_2 y_m + (p_5 u + p_3)(p_3 + p_4)x_2 - p_5 x_2 \frac{du}{dt} = 0.
$$

Multiplying this equation by $(p_5 u + p_3)$ and replacing all terms in $(p_5 u + p_3)x_2$ by their expression taken from the first component of the state equation, we get the input/output differential equation

$$
(p_5 u + p_3)\frac{d^2 y_m}{dt^2} + \left[(p_5 u + p_3)(p_1 + p_2 + p_3 + p_4) - p_5 \frac{du}{dt} \right]\frac{dy_m}{dt}
$$

$$
+ \left\{ u[p_5(p_1 + p_2)(p_3 + p_4) - 2p_5 p_2 p_3] - u^2(p_5^2 p_2) - p_5(p_1 + p_2)\frac{du}{dt} \right.
$$

$$
\left. + p_3(p_1 p_3 + p_1 p_4 + p_2 p_4) \right\} y_m = 0,
$$

which we shall divide (for instance) by p_5 to avoid having an infinite number of ways of writing down the same equation. The initial conditions remain to be computed as

$$
y_m(0) = x_1(0) = 1,
$$

$$
\frac{dy_m}{dt}(0) = \frac{dx_1}{dt}(0) = -(p_1 + p_2)x_1(0) + (p_5 u(0) + p_3)x_2(0) = -(p_1 + p_2).
$$

A *sufficient* condition for $M(\hat{\boldsymbol{p}}) = M(\boldsymbol{p}^*)$ to be satisfied is that all coefficients of the input/output differential equation (normalized by division by p_5) and initial conditions be identical. This condition is equivalent to the set of equations

$$\frac{\hat{p}_3}{\hat{p}_5} = \frac{p_3^*}{p_5^*}, \qquad\qquad\qquad \left(\text{term in } \frac{\mathrm{d}^2 y_m}{\mathrm{d}t^2}\right)$$

$$\hat{p}_1 + \hat{p}_2 + \hat{p}_3 + \hat{p}_4 = p_1^* + p_2^* + p_3^* + p_4^*, \qquad\qquad \left(\text{term in } \frac{\mathrm{d}y_m}{\mathrm{d}t}u\right)$$

$$\frac{\hat{p}_3}{\hat{p}_5}(\hat{p}_1 + \hat{p}_2 + \hat{p}_3 + \hat{p}_4) = \frac{p_3^*}{p_5^*}(p_1^* + p_2^* + p_3^* + p_4^*), \qquad \left(\text{term in } \frac{\mathrm{d}y_m}{\mathrm{d}t}\right)$$

$$\hat{p}_2 \hat{p}_5 = p_2^* p_5^*, \qquad\qquad\qquad \left(\text{term in } y_m u^2\right)$$

$$\hat{p}_1 + \hat{p}_2 = p_1^* + p_2^*, \qquad\qquad\qquad \left(\text{term in } y_m \frac{\mathrm{d}u}{\mathrm{d}t}\right)$$

$$(\hat{p}_1 + \hat{p}_2)(\hat{p}_3 + \hat{p}_4) - 2\hat{p}_2\hat{p}_3 = (p_1^* + p_2^*)(p_3^* + p_4^*) - 2p_2^* p_3^*, \quad (\text{term in } y_m u)$$

$$\frac{\hat{p}_3}{\hat{p}_5}(\hat{p}_1\hat{p}_3 + \hat{p}_1\hat{p}_4 + \hat{p}_2\hat{p}_4) = \frac{p_3^*}{p_5^*}(p_1^* p_3^* + p_1^* p_4^* + p_2^* p_4^*), \qquad (\text{term in } y_m)$$

$$\hat{p}_1 + \hat{p}_2 = p_1^* + p_2^*. \qquad\qquad\qquad\qquad (\text{initial conditions})$$

Four of these equations can be deduced from the four others, so that there are less independent equations than there are unknowns. This set of equations therefore possesses an uncountable set of solutions for $\hat{p}$ (including the trivial solution $\hat{p} = p^*$), so that $M(.)$ is s.n.i.

Remark 5
(i) This technique makes it possible to handle model structures that are defined by sets of differential equations relating inputs, outputs and internal (or latent) variables which are not necessarily state variables. It is thus not limited to state-space models.

(ii) Eliminating state variables (or more generally latent variables) raises nontrivial difficulties not discussed here (Diop, 1989). The result sometimes consists of a set of differential equations ($= 0$) and non-equations ($\neq 0$) binding inputs and outputs. Moreover, the expressions obtained are often relatively complicated, and it is difficult to know whether they could be simplified, which would modify the equations to be taken into account to ensure that $M(\hat{p}) = M(p^*)$.

(iii) If this approach is to be used in order to prove that a model is s.g.i., one should make sure that the input/output differential equations thus obtained are in reduced form (i.e. no simplification is possible) and that the initial conditions preserve the generic nature of their solution (i.e. it is possible to estimate all coefficients appearing in these equations from ideal input/output data). For more details, see Ollivier (1990).

Finally, let us consider a model defined by a set of relations of the type

$$r_k(\boldsymbol{y}_m(t), \boldsymbol{u}(t), \boldsymbol{x}(t), \boldsymbol{p}) = 0, k = 1, \ldots, n_r$$

where the r_k are polynomial functions of the inputs u, outputs y_m, internal variables x and their derivatives with respect to time. Glad and Ljung (1990) have shown that if a parameter p_k of this model is s.g.i., then it can be estimated (at least in principle and within the idealized framework of identifiability studies) by solving an affine equation of the type

$$a(y_m(t), u(t))p_k = b(y_m(t), u(t)),$$

where a and b are polynomial functions of the inputs, outputs and their derivatives with respect to time (see also Ljung and Glad (1994)). They provide a method to obtain these functions, which constitutes a test of global identifiability. The condition $a(y_m(t), u(t)) \neq 0$ indicates the type of input that must be used to allow the identification to take place. It is analogous to the condition of persistent excitation, well known in the linear case.

3.7   Connexions with experimental design

The methods presented so far make it possible to assess the structural identifiability and structural distinguishability of the models considered. If serious defects are detected, it is possible to determine which interactions with the system (via additional sensors or actuators) might remove the ambiguity. One may thus talk of *qualitative experimental design*. One should then move from qualitative to quantitative and look for the experiment that can be expected to yield the best possible data among all the feasible experiments. It is indeed well known that the quality of a model depends heavily on the quality of the data used for its elaboration. *Quantitative experimental design* has given rise to considerable literature (see, e.g., the books by Fedorov (1972), Zarrop (1979), Silvey (1980), Pazman (1986) and Atkinson and Donev (1992) and the papers by Titterington (1980), Steinberg and Hunter (1984) and Walter and Pronzato (1990)). We shall limit ourselves here to a very succinct presentation of the subject. Designing an experiment may, for example, mean choosing the inputs, the measurement times, and the sensors and actuators characteristics and locations. We shall assume that any experiment is characterized by a vector e. If the problem is, for instance, to choose measurement times, the entries of e will be the times at which these measurements will have to be performed. The optimal experiment will obviously depend on the objectives to be achieved (parameter estimation, structure discrimination, optimal control, etc.). We shall assume here that the objective is to estimate p with the best possible precision, which may be viewed as maximizing some measure of identifiability. One could also assume that the objective is to best discriminate between several candidate model structures (see, e.g., Atkinson and Cox, 1974; Hill, 1978; Huang, 1991), which could be viewed as the maximization of some measure of distinguishability. Choosing e requires being able to predict, for any feasible experiment, the resulting uncertainty on the estimate of p. One must therefore have at one's disposal a model of the noise corrupting the data as well as a method to characterize uncertainties, simple enough to allow optimization. We shall assume that

$$y(e) = y_m(p^*, e) + \varepsilon^*(e),$$

where $y(e)$ is the N-dimensional vector containing all measurements that will be available on the process once the experiment e has been performed, y_m is the corresponding model output, p^* is the true value for the parameters and ε^* is the vector of the measurement noises, assumed to have zero mean and known covariance $\Sigma(e)$. (One could also consider other noise configurations by replacing the output error by a suitable prediction error.) Under technical regularity conditions, provided that $M(.)$ is s.g.i. and e is sufficiently well chosen, the maximum-likelihood estimator of p from $y(e)$ tends, when N tends to infinity, to be distributed according to a Gaussian law with mean p^* and covariance $\frac{1}{N}M_F^{-1}(p^*, e)$, where M_F is the normalized Fisher information matrix, given by

$$M_F(p, e) = \frac{1}{N}X^T(p, e)\Sigma^{-1}(e)X(p, e)$$

with

$$X(p, e) = \frac{\partial y_m(p, e)}{\partial p^T}.$$

The Fisher information matrix is therefore particularly easy to compute, since it only involves the sensitivity of the model outputs with respect to the parameters, and the noise covariance. The larger M_F is (in a sense to be specified) the less scattered the estimates will be around the true value p^*. For these two reasons, almost all criteria used to design experiments for parameter estimation are functionals of the Fisher information matrix. Thus, an experiment is D-optimal if it maximizes the criterion

$$J(e) = \det M_F(p, e),$$

which corresponds to minimizing the volume of the asymptotic confidence ellipsoids on the parameters. D-optimal design is only possible if $M(.)$ is at least s.l.i. Otherwise the determinant of the Fisher information matrix will be identically zero whatever the experiment (which provides a numerical means to detect that the model is not locally identifiable). Except when the model is LP, M_F depends on p, which is of course not known *a priori*. Various techniques may be used to eliminate this dependency in p while taking advantage of prior information. One may, for instance, use a nominal value p_0 for the parameters, or look for the experiment that is best on average over all possible values for p, or the best for the worst possible value of p (see (Walter and Pronzato, 1990)). The problem then reduces to one of constrained nonlinear optimization, for which the tools of nonlinear programming can be used.

3.8 Bibliography

[1] ANDERSON, D. H. (1982). *Compartmental Modeling and Tracer Kinetics*, Springer, Berlin.

[2] ÅSTRÖM, K. J. (1972). System Identification. *Lecture Notes Math. Germ.*, 294, 35–55.

[3] ATKINSON, A. C. AND D. R. COX (1974). Planning experiments for discriminating between models (with discussion), *J. Royal. Statist. Soc.*, B 36, 321–348.

[4] ATKINSON, A. C. AND A. N. DONEV (1992). *Optimal Experimental Design*, Oxford University Press, Oxford.

[5] AUDOLY, S. AND L. D'ANGIO (1983). On the identifiability of linear compartmental systems: a revisited transfer function approach based on topological properties. *Math. Biosci.*, 66, 201–228.

[6] BELLMAN, R. AND K. J. ÅSTRÖM (1970). On structural identifiability. *Math. Biosci.*, 7, 329–339.

[7] BERMAN, M. AND R. SCHOENFELD (1956). Invariants in experimental data on linear kinetics and the formulation of models. *J. of Applied Physics*, 27, 1361–1370.

[8] BERNTSEN, H. E. AND J. G. BALCHEN (1973). *Identifiability of linear dynamical systems*. Proc 3rd IFAC Symp. on Identification and System Parameter Estimation, The Hague, 871–874.

[9] BUCHBERGER, B. (1970). Ein algorithmisches Kriterium für die Lösbarkeit eines algebraischen Gleichungssystems. *Aeq. Math.*, 4, 374–383.

[10] CHAPPELL, M. J., K. R. GODFREY AND S. VAJDA (1990). Global identifiability of the parameters of nonlinear systems with specified inputs: a comparison of methods. *Math. Biosci.*, 102, 41–73.

[11] COBELLI, C., A. LEPSCHY AND G. ROMANIN-JACUR (1979a). Identifiability results on some constrained compartmental systems. *Math. Biosci.*, 47, 173–196.

[12] COBELLI, C., A. LEPSCHY AND G. ROMANIN-JACUR (1979b). Identifiability of compartmental models and related structural properties. *Math. Biosci.*, 48, 1–18.

[13] DIOP, S. (1989). *Théorie de l'élimination et principe du modèle interne en automatique*. Doctorate thesis, Université Paris Sud, Orsay, France.

[14] DIOP, S. AND M. FLIESS (1991). *Nonlinear observability, identifiability and persistent trajectories*. Proc. 30th IEEE CDC, Brighton, 714–719.

[15] FEDOROV, V. V. (1972). *Theory of Optimal Experiments*. Academic Press, New York.

[16] FISHER, F. M. (1966). *The Identification Problem in Economics*. McGraw-Hill, New York.

[17] FLIESS, M. (1981). Fonctionnelles causales non linéaires et indéterminées non commutatives. *Bull. Soc. Math. de France.* 109, 3–40.

[18] FLIESS, M. (1989). Automatique et corps différentiels. *Forum Math.*, 1, 227–238.

[19] GLAD, S. T. AND L. LJUNG (1990). *Parametrization of nonlinear model structures as linear regressions*. Prep. 11th IFAC World Congress, Tallinn, 6, 67–71.

[20] GLOVER, K. AND J. C. WILLEMS (1974). Parametrizations of linear dynamical systems: canonical forms and identifiability. *IEEE Trans. Autom. Control*, AC-19, 640–644.

[21] GREWAL, M. S. AND K. GLOVER (1976). Identifiability of linear and nonlinear dynamical systems. *IEEE Trans. Autom. Control*, AC-21, 833–837.

[22] HERMANN, R. AND A. J. KRENER (1977). Nonlinear controllability and observability. *IEEE Trans. Autom. Control*, AC-22, 728–740.

[23] HILL, P. D. H. (1978). A review of experimental design procedures for regression model discrimination, *Technometrics*, 20, 15–21.

[24] HORST, R. AND H. TUY (1990). *Global Optimization, Deterministic Approaches*. Springer-Verlag, Berlin.

[25] HUANG, C.-Y. (1991). *Planification d'expériences pour la discrimination entre modèles*. Doctorate thesis, Université de Paris Sud, Orsay, France.

[26] LAUNAY, G. (1989). *Modélisation du stockage de la sérotonine dans les plaquettes humaines*. Doctorate thesis, Université Paris Dauphine, Paris, France.

[27] LECOURTIER, Y., F. LAMNABHI-LAGARRIGUE AND E. WALTER (1987). *Volterra and generating power series approaches to identifiability testing*. In E. Walter (ed.), Identifiability of Parametric Models. Pergamon, Oxford, pp. 50–66.

[28] LECOURTIER, Y. AND A. RAKSANYI (1987). *The testing of structural properties through symbolic computation*. In E. Walter (ed.), Identifiability of Parametric Models. Pergamon, Oxford, pp. 75–84.

[29] LJUNG, L. (1987). *System Identification, Theory for the User*. Prentice-Hall, Englewood Cliffs.

[30] LJUNG, L. AND T. GLAD (1994). On global identifiability for arbitrary model parametrizations. *Automatica*, 30 (2), 265–276.

[31] NORTON, J. P. (1986). *An Introduction to Identification*. Academic Press, London.

[32] OLLIVIER, F. (1990). *Le problème de l'identifiabilité structurelle globale: approche théorique, méthodes effectives et bornes de complexité*. Doctorate thesis, Ecole Polytechnique, Palaiseau, France.

[33] PAZMAN, A. (1986). *Foundations of Optimum Experimental Design*. VEDA, Bratislava.

[34] POHJANPALO, H. (1978). System identifiability based on the power series expansion of the solution, *Math. Biosci.*, 41, 21–33.

[35] RATSCHEK, H. AND J. ROKNE (1988). *New Computer Methods for Global Optimization*. Ellis Horwood, Chichester. (Distributed by John Wiley).

[36] RITT, J. F. (1950). *Differential Algebra*, American Mathematical Society, Providence.

[37] SILVEY, S. D. (1980). *Optimal Design*. Chapman & Hall, London.

[38] STEINBERG, D. M. AND W. G. HUNTER (1984). Experimental design: review and comment (with discussion). *Technometrics*, 26, 71–130.

[39] SUSSMAN, H. J. (1977). Existence and uniqueness of minimal realizations of non-linear systems. *Math. Syst. Theory*, 10, 263–284.

[40] TITTERINGTON, D. M. (1980). Aspects of optimal design in dynamic systems. *Technometrics*, 22, 287–299.

[41] TUNALI, E. T. AND T. J. TARN (1987). New results for identifiability of non-linear systems. *IEEE Trans. Autom. Control*, AC-32, 146–154.

[42] VAJDA, S., K. R. GODFREY AND H. RABITZ (1989). Similarity transformation approach to identifiability analysis of nonlinear compartmental models. *Math. Biosci.*, 93, 217–248.

[43] VAJDA, S. AND H. RABITZ (1989). State isomorphism approach to global identifiability of nonlinear systems. *IEEE Trans. Autom. Control*, AC-34, 220–223.

[44] VAJDA, S., H. RABITZ, E. WALTER AND Y. LECOURTIER (1989). Qualitative and quantitative identifiability analysis of nonlinear chemical models. *Chem. Eng. Comm.*, 83, 191–219.

[45] WALTER, E. (1982). *Identifiability of State Space Models*. Springer, Berlin.

[46] WALTER, E. (ED.) (1987). *Identifiability of Parametric Models*. Pergamon, Oxford.

[47] WALTER, E. AND Y. LECOURTIER (1981). Unidentifiable compartmental models: what to do? *Math. Biosci.*, 56, 1–25.

[48] WALTER, E. AND Y. LECOURTIER (1982). Global approaches to identifiability testing for linear and nonlinear state space models. *Math. and Comput. in Simul.*, 24, 472–482.

[49] WALTER, E., Y. LECOURTIER AND J. HAPPEL (1984). On the structural output distinguishability of parametric models, and its relations with structural identifiability. *IEEE Trans. Autom. Control*, AC- 29, 56–57.

[50] WALTER, E., Y. LECOURTIER, A. RAKSANYI AND J. HAPPEL (1985). On the distinguishability of parametric models with different structures. In J. Eisenfeld and C. De Lisi (eds), *Mathematics and Computers in Biomedical Applications*, North-Holland, Amsterdam, pp. 145–160.

[51] WALTER, E., H. PIET-LAHANIER AND J. HAPPEL (1986). Estimation of non-uniquely identifiable parameters via exhaustive modeling and membership set theory, *Math. and Comput. in Simul.*, 28, 479–490.

[52] WALTER, E. AND L. PRONZATO (1990). Qualitative and quantitative experiment design for phenomenological models - a survey. *Automatica*, 26, 195–213.

[53] WALTER, E. AND L. PRONZATO (1994). *Identification de modèles paramétriques à partir de données expérimentales*. Masson, Paris.

[54] ZARROP, M. B. (1979). *Optimal Experiment Design for Dynamic System Identification*. Springer, Heidelberg.

CHAPTER 4

Identification and realization by state affine models

J. LOTTIN, D. THOMASSET

4.1 Introduction

Representation of a "black-box" type, an alternative to the knowledge of the theoretical model, is well mastered in the cases of stationary linear systems. Indeed, by means of experiments carried out on the process, many identification procedures lead to state space equations or a transfer matrix representation. In the varylinear case, the model depends upon one or several influential parameters. It is interesting to obtain a global model, that is, a model which validly represents the system in the whole working range, whether for simulation or control law synthesis. Determining the exact dependence laws of model coefficients with respect to this or these influential parameters is tricky. The problem may be simplified if one is interested in regulation around a finite number of setting points. The procedure that is described in the next paragraphs shows how to get such a model by the use of state affine representation. After setting the retained hypothesis in section 4.2 and the mathematical background in section 4.3, two approaches to the identification problem are presented in section 4.4. One is a methodology for realization from a series of identification of the "linear tangent" around each setting point. The other is characterized by a direct identification. Both methods lead to obtaining a nonlinear discrete time global model (state affine model). Various applications of these techniques have been developed in close collaboration with the industrial world. Two examples are presented in section 4.5: The first concerns the modeling of a helicopter and the second deals with identification of a chemical neutralization process. These application results have justified the development of the AFFINE (software for Personal Computer), that uses Matlab facilities. This tool which is presented in section 4.6, allows a representation of varylinear systems. A particular application of state affine models which concerns the representation of a process having an asymmetry of gain around equilibrium points is described in section 4.7. Finally, section 4.8 extends the method to the MIMO systems. The additional problem to be solved concerns the minimality of the realization.

4.2 Hypothesis

In this part, we focus on continuous systems whose dynamical behavior is in an implicit or explicit manner function of a measurable influential parameter, denoted q and called "variable factor". This factor may evolve in a range $[q_{min}, q_{max}]$ which characterizes the finite set of operating points of the process. The dynamical behavior of such a system is analyzed considering the factor q as a constant or slow variable. Other more general studies attempt to cast off this restrictive assumption (Bornard and Hu, 1987). In order to respect the hypothesis of the approximation theorem (Fliess and Normand-Cyrot, 1980) which justifies the use of state affine models, we consider that the current process output continuously depends upon past output samples. Furthermore it is assumed that for a given value of the factor q, the system is linear with respect to control inputs and that dependence laws of model coefficients are continuous functions of this factor q.

4.3 State affine models

4.3.1 Definition

In discrete time, state affine models are linear systems with respect to state variables and have the following general form:

$$\begin{cases} x_{k+1} = f(u_k)x_k + g(u_k) \\ y_k = Cx_k \end{cases} \tag{4.1}$$

where $x_k \in R^n, u_k \in R^m, y_k \in R^p$ f_{ij}, g_j designate respectively matrix f and vector g components and are analytic functions of R^m in R. In fact, we retain the subclass of systems with polynomial inputs whose state representation is given by:

$$x_{k+1} = \left[A_o + \sum_j {}_\tau P_j (u_{1_k} \ldots u_{m_k}) A_j \right] x_k \tag{4.2}$$

$$y_k = Cx_k$$

where ${}_\tau P_j$ is a monomial of degree less than or equal to τ with respect to $u_{1_k} \ldots u_{m_k}$ and A_o, A_j are $R^n \longrightarrow R^n$ linear. For example, the case $m = 3, \tau = 2$ leads to the following expansion:

$$x_{k+1} = \left[A_o + u_{1_k} \cdot A_1 + u_{2_k} \cdot A_2 + u_{3_k} \cdot A_3 + u_{1_k}^2 \cdot A_4 + u_{2_k}^2 \cdot A_5 \right.$$
$$\left. + u_{3_k}^2 \cdot A_6 + u_{1_k} u_{2_k} A_7 + u_{2_k} u_{3_k} A_8 + u_{1_k} u_{3_k} A_9 \right] \cdot x_k \tag{4.3}$$

$$y_k = Cx_k$$

This choice is motivated by two main reasons:

- in the continuous case, the class of bilinear systems allows us to represent, at least in an approximate way, many nonlinear systems. Unfortunately, in the discrete time case, this kind of model is not as rich, and discretization of a continuous bilinear model introduces in a natural way products of inputs at the same time and leads to the use of state affine models (Monaco and Normand-Cyrot, 1987).

– Furthermore, one knows the existence of a theorem which states that every system in discrete time over R, whose outputs depend continuously upon inputs, can be approximated by a finite dimension state affine model in a compact time interval and for all bounded inputs (Fliess and Normand-Cyrot, 1980).

This theorem proves the existence of an approximation. It does not provide a construction algorithm but justifies the search for state affine models that approach a nonlinear dynamical behavior and *a fortiori* varylinear systems. There is no unicity of the obtained realization. From this point of view, the state affine model should be considered as a discrete approximation of the initial system. The number of retained monomials will result from a compromise between the model complexity and the quality of the approximation.

4.3.2 Application to varylinear systems

Let us now consider varylinear systems (Thomasset and Neyran, 1985; Dufour and Thomasset, 1984; Thomasset, 1987). It means that dynamical behavior is a function of external influential parameters and as long as they keep a constant value, the system behavior is linear. Therefore, the case of only one influential parameter denoted q is treated. The process can then be considered as a nonlinear system with an extended input vector composed with control variables, u, and the influential parameter q, which leads to the search for a state affine model. However it should be noted that only the dynamics with respect to control variables u is studied and not the dynamics with respect to parameter q. In order to preserve the linearity feature of the model when q is constant, the polynomial expansions concern only the parameter q. This is why the state affine representation has the following form in the case of only one input variable u:

$$x_{k+1} = \left[\sum_{j=0}^{\sigma_1} q^j(k) \cdot A_j + \sum_{j=0}^{\sigma_2} u_k \cdot q^j(k) \cdot B_j \right] \cdot x_k$$

$$y_k = C \cdot x_k \tag{4.4}$$

It is possible to give an interpretation of A_i and B_j matrices. Indeed let us consider the discrete state space representation of the varylinear system:

$$\xi_{k+1} = \alpha(q) \cdot \xi_k + \beta(q) \cdot u_k$$

$$y_k = \gamma \cdot \xi_k \tag{4.5}$$

where $\alpha(q)$ and $\beta(q)$ matrices are not necessarily specified but are assumed to be approximated by the following polynomial expansions:

$$\alpha(q) = \sum_{j=0}^{\sigma_1} q^j \cdot A_j^*$$

$$\beta(q) = \sum_{j=0}^{\sigma_2} q^j \cdot B_j^* \tag{4.6}$$

The state affine (4.4) model is obtained when considering the extended state vector:

$$x_k = \begin{bmatrix} \xi_k \\ 1 \end{bmatrix} \tag{4.7}$$

and the matrix relations:

$$A_0 = \begin{bmatrix} A_0^* & 0 \\ 0 & 1 \end{bmatrix}; \quad A_j = \begin{bmatrix} A_j^* & 0 \\ 0 & 0 \end{bmatrix}; \quad B_j = \begin{bmatrix} 0 & B_j^* \\ 0 & 0 \end{bmatrix}; C = [\gamma \ 0] \qquad (4.8)$$
$$\phantom{A_0 = \begin{bmatrix} A_0^* & 0 \\ 0 & 1 \end{bmatrix}; \quad} j \neq 0$$

Laws of dependence with respect to q of matrices $\alpha(q)$ and $\beta(q)$ are rarely explicitly known. Indeed matrices $\alpha(q)$ and $\beta(q)$ result from the discretization of a continuous process and when relations are known in continuous time, it is difficult to derive them in discrete time, as shown in the following example which concerns a second-order process whose poles are solutions of the equation:

$$s^2 + 2\lambda(q)\omega_n s + \omega_n^2 = 0.$$

It means that the sole damping factor depends upon the influential parameter q. The poles of the discretized system with a sampling period T are therefore solutions of the equation:

$$z^2 - 2z\,e^{(-\lambda(q)\omega_n T)} \cos(\omega_n T(1 - \lambda^2(q))^{0.5}) + e^{-2\lambda(q)\omega_n T}) = 0.$$

Even if the form of the function $\lambda(q)$ is known, it is not easy to find its coefficients from the knowledge of the poles in discrete time for a few values of q. This is why it is preferred to search for an approximated model, of known structure, knowing that it will be able to represent as closely as possible the true process. Two methods are proposed in order to obtain the state affine model:

- The indirect approach goes through the estimation of $\alpha(q)$ and $\beta(q)$ matrices for some known values of the parameter q.
- The direct approach uses an identification sequence in which the control variable evolves with a fast dynamic, while the influential parameter evolves slowly in order to cover its variation domain.

4.4 Methodology: modeling and identification of a varylinear system by a state affine model

4.4.1 Indirect approach

This approach is called indirect since it comprises three distinct steps:

- Identification carried out by means of a classical method, around N_e equilibrium points that are parametrized by the measurable factor q, of a family of N_e stationary linear models.
- Construction of a bilinear realization for each model.
- Construction of the global state affine model by matricial regression.

It is therefore required that all the state space models of the family are known in the same basis. The problem of the choice of the ideal basis is then questioned, or in other words the choice of the state space representation to be used. Wilkinson mentioned the problem of sensitivity of location of the roots of a polynomial with respect to its coefficients. It seems preferable to operate the regressions on the eigenvalues by choosing a diagonal form of the state matrix, rather than on the coefficients of the characteristic polynomial by choosing a companion form. However, this raises the following questions:

- the diagonal form is difficult to use when eigenvalues are multiple or complex conjugate, since it is then necessary to change the structure;
- even in the case of real eigenvalues, which can be examined on the family of models that are obtained, these evolve in a discontinuous way with respect to the external parameter.

Let us illustrate this sensitivity problem with a third-order example: Suppose that the time constants of a linear continuous system are given by the following relations, for q varying from 2 to 6:

$$\tau_1 = 10 - q$$
$$\tau_2 = 2(q-4)^2 + 4$$
$$\tau_3 = 3 + 2(q-2)$$

Figure 4.1 shows the evolution of the discrete poles for a set of values of q when:

- Exact discretization is carried out: fig. 4.1a.
- For each pole, a polynomial regression of order 5 is carried out from the values obtained for the set of q, then poles are computed from these approximations, for the same set of values of q in order to make the comparison: fig. 4.1b.
- First, the poles that are computed for each value of q are classified in decreasing order, then a polynomial regression is carried out and poles are computed as above (there is not the same continuity as for initial poles): fig. 4.1c.
- For each value of q, coefficients of the characteristic polynomial are computed, then a regression of order 5 is carried out on polynomial coefficients and, for each value of q, the roots of this new polynomial are computed: fig. 4.1d.

This example shows that the result obtained from the characteristic polynomial (figure 4.1d) is at least as good as the one obtained from the classification of the roots (figure 4.1c). Note that the case of figure 4.1b does not correspond to a realistic situation since no information about the good classification is known. This is why the companion form was retained (minimal number of coefficients over which the polynomial regression is carried out and no problem of structure change). A second problem to be solved concerns the choice of the orders of the polynomial expansions. The compromise between the complexity of calculations and the quality of the model obtained has been already mentioned, and this should be connected to the quality of the linear models that are used for regression. Rather than choosing different orders for each coefficient of the representation, only two orders are searched; one (σ_1) concerns the coefficients of the state matrix, the other (σ_2) is attached to the coefficients of the input matrix. A criterion of Akaike's type (Ljung, 1987) may be used to evaluate an acceptable value of σ_1 and σ_2 orders. The main disadvantage of the indirect approach is consideration of particular values of the parameter q to build the linear models. These are arbitrarily chosen, which may induce a lack of information about the dependence with respect to the external parameter in some working domains.

4.4.2   Direct approach

In order to avoid these problems, this section proposes a more direct method which allows state affine model coefficients to be determined from a sequence of inputs and outputs wich is obtained by exciting the control input while the parameter q varies in the range

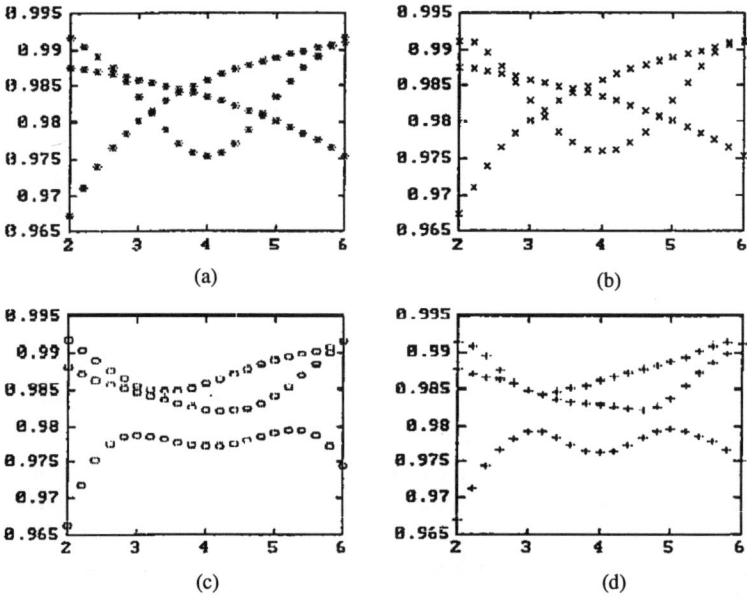

Figure 4.1: Variations of discretized poles.

(q_{min}, q_{max}). A single input single output linear system can be described by a canonical companion form. The one leading to the simplified form of the output matrix is chosen, since it is independent of variations of the parameter q.

$$
\begin{bmatrix} \xi_{1k+1} \\ \cdot \\ \cdot \\ \cdot \\ \cdot \\ \xi_{nk+1} \end{bmatrix} = \begin{bmatrix} \alpha_1 & 1 & 0 & . & . & . & . & 0 \\ \cdot & & & & & & & \cdot \\ \cdot & & & & & & & \cdot \\ \cdot & & & & & & & \cdot \\ \cdot & & & & & & . & 1 \\ \alpha_n & 0 & . & . & . & . & . & 0 \end{bmatrix} \begin{bmatrix} \xi_{1k} \\ \cdot \\ \cdot \\ \cdot \\ \cdot \\ \xi_{nk} \end{bmatrix} + \begin{bmatrix} \beta_1 \\ \cdot \\ \cdot \\ \cdot \\ \cdot \\ \beta_n \end{bmatrix} u_k \qquad (4.9)
$$

$$
y_k = [10\ldots0] \begin{bmatrix} \xi_{1k} \\ \cdot \\ \cdot \\ \cdot \\ \xi_{nk} \end{bmatrix} \qquad (4.10)
$$

that is

$$
\xi_{k+1} = \alpha\xi_k + \beta u_k
$$
$$
y_k = \gamma\xi_k
$$

For the above mentioned reasons, it is that form which is retained for the state space representation. In the case of invariant processes, α and β matrices are constant. When the process depends upon a parameter q, then α and β matrices depend on q and polynomial expansions of $\alpha(q)$ and $\beta(q)$ matrices are searched with the form:

$$
\begin{aligned}
\alpha(q) &= A_0^* + qA_1^* + \cdots + q^{\sigma_1}A_{\sigma_1}^* \\
\beta(q) &= B_0^* + qB_1^* + \cdots + q^{\sigma_2}B_{\sigma_2}^*
\end{aligned}
\tag{4.11}
$$

with

$$
\alpha = \begin{bmatrix}
a_{10} & 1 & 0 & . & . & 0 \\
 & 0 & 1 & . & . & 0 \\
. & & & & & . \\
. & & & & & . \\
. & & 0 & & & 1 \\
a_{n0} & 0 & . & . & . & 0
\end{bmatrix}
+ q. \begin{bmatrix}
a_{11} & 0 & . & . & 0 \\
. & & & & . \\
. & & & & . \\
. & & & & . \\
. & & & & . \\
a_{n1} & 0 & . & . & 0
\end{bmatrix}
+ \cdots
$$

$$
+ q^{\sigma_1} \begin{bmatrix}
a_{1\sigma_1} & 0 & . & . & 0 \\
. & & & & . \\
. & & & & . \\
. & & & & . \\
. & & & & . \\
a_{n\sigma_1} & . & . & . & 0
\end{bmatrix}
\tag{4.12}
$$

$$
\beta = \begin{bmatrix} b_{10} \\ . \\ . \\ . \\ b_{n0} \end{bmatrix}
+ q. \begin{bmatrix} b_{11} \\ . \\ . \\ . \\ b_{n1} \end{bmatrix}
+ \cdots + q^{\sigma_2} \begin{bmatrix} b_{1\sigma_2} \\ . \\ . \\ . \\ b_{n\sigma_2} \end{bmatrix}
$$

The difference equation which links $y(k)$ to the input, output, and parameter past samples is easily obtained (since q varies with time, $q(k)$ means the value of q at time k);

$$
\begin{aligned}
y_k &= a_{10}y_{k-1} + a_{11}y_{k-1}q(k-1) + \ldots + a_{1\sigma_1}y_{k-1}\, q^{\sigma_1}(k-1) \\
&\quad + b_{10}u_{k-1} + b_{11}u_{k-1}q(k-1) + \ldots + b_{1\sigma_2}u_{k-1}\, q^{\sigma_2}(k-1) \\
&\quad + a_{20}y_{k-2} + a_{21}y_{k-2}q(k-2) + \ldots + a_{2\sigma_1}y_{k-2}\, q^{\sigma_1}(k-2) \\
&\quad + \cdots\cdots\cdots \\
&\quad + a_{n0}y_{k-n} + a_{n1}y_{k-n}q(k-n) + \ldots + a_{n\sigma_1}y_{k-n}\, q^{\sigma_1}(k-n) \\
&\quad + b_{n0}u_{k-n} + b_{n1}u_{k-n}q(k-n) + \ldots + b_{n\sigma_2}u_{k-n}\, q^{\sigma_2}(k-n)
\end{aligned}
\tag{4.13}
$$

This model is linear with respect to unknown parameters $\theta = [a_{10}, \ldots, b_{n\sigma_2}]^T$. An explicit estimator in the least square sense of the vector of unknowns can be constructed. But first, it is meaningful to study the influence of white noise that is added to the output:

$$
y_k^b = y_k + w_k.
$$

The difference equation which gives the output is then modified as follows:

$$
y_k^b = \sum_{j=0}^{\sigma_1}\sum_{i=1}^{n} a_{ij}q^j(k-i)y_{k-i}^b + \sum_{j=0}^{\sigma_2}\sum_{i=1}^{n} b_{ij}q^j(k-i)u_{k-i} + E_k
\tag{4.14}
$$

where $E(k)$ is a correlated residue with the following expression:

$$
\begin{aligned}
E_k \;=\;\; & w_k - a_{10}w_{k-1} - a_{20}w_{k-2} - \cdots - a_{n0}w_{k-n} \\
& - a_{11}w_{k-1}q(k-1) - a_{21}w_{k-2}q(k-2) - \cdots - a_{n1}w_{k-n}q(k-n) \\
& \cdots \\
& - a_{1\sigma_1}w_{k-1}q^{\sigma_1}(k-1) - a_{2\sigma_1}w_{k-2}q^{\sigma_1}(k-2) - \cdots \\
& \quad - a_{n\sigma_1}w_{k-n}q^{\sigma_1}(k-n)
\end{aligned}
$$

$$
= w_k - \sum_{j=0}^{\sigma_1}\sum_{i=1}^{n} a_{ij}q^j(k-i)w_{k-i}. \tag{4.15}
$$

An optimal solution in the least square sense is searched. In order to do so, the difference equation of y_k is written for $m \le k \le m + N$, with $m > n$, which leads to the matricial equation:

$$
Y = X\theta + E \tag{4.16}
$$

in which:

$$
\begin{aligned}
Y &= [y_m^b....y_{m+N}^b]^T \\
E &= [E_m......E_{m+N}]^T \\
\theta &= [a_{10}, a_{11}, \ldots, a_{1\sigma_1}, a_{20}, a_{21}, \ldots, a_{n\sigma_1}, b_{10}, b_{11} \ldots, b_{1\sigma_2}, b_{20}, b_{21} \ldots, b_{n\sigma_2}]^T
\end{aligned}
\tag{4.17}
$$

and X is a matrix with dimensions $((N+1), n(\sigma_1+1)+n(\sigma_2+1))$, that can be decomposed in $[X_y, X_u]$ with X_y and X_u of respective dimensions $((N+1), n(\sigma_1+1))$ and $((N+1), n(\sigma_2+1))$. The matrix X_y is itself structured in n submatrices of dimensions $((N+1), (\sigma_1+1))$, the j^{ith} submatrix being given by:

$$
\begin{bmatrix}
y_{m-j}^b, y_{m-j}^b q(m-j), \ldots\ldots\ldots\ldots, y_{m-j}^b q^{\sigma_1}(m-j) \\
y_{m-j+1}^b, y_{m-j+1}^b q(mj+1), \ldots\ldots, y_{m-j+1}^b q^{\sigma_1}(m-j+1) \\
\quad .. \\
\quad .. \\
\quad .. \\
y_{m-j+N}^b, y_{m-j+N}^b q(m-j+N), \ldots.., y_{m-j+N}^b q^{\sigma_1}(m-j+N)
\end{bmatrix}
\tag{4.18}
$$

The matrix X_u is built in the same way when replacing y^b by u and σ_1 by σ_2 The estimation of θ in the unweighted least square sense is given by the well-known formula:

$$
\hat{\theta} = (X^T X)^{-1} X^T Y. \tag{4.19}
$$

This estimator is unbiased if on the one hand the residue E is centered, which is the case from the assumption that the noise b has zero mean, and on the other hand there is no correlation between X and E matrices. This last condition is not satisfied because the parameters a_{ij} which appear in the expression of the residue (4.15) are correlated with the coefficients of the X matrix. One can try to filter the residue in order to obtain an unbiased estimator. Therefore, two methods are presented. The first one is a direct application of Clarke's algorithm to the generalized least square method (Clarke, 1967; Mouille, 1990). The second one takes into account the varylinear feature of the system to be identified, to derive a varylinear filter (Thomasset, 1987; Mouille, 1990). Whichever method is used, the identification proceeds as follows:

(1) excitation of the control input, for example by means of a pseudo random binary sequence, with simultaneous, but much slower variation, of the external parameter q,

(2) acquisition of control u, parameter q and output y sequences,

(3) construction of X and Y matrices,

(4) estimation of coefficient vector $\hat{\theta}$,

(5) calculation of the output sequence of the obtained model, when submitted to the same entries (control u and parameter q),

(6) calculation of the mean square error between the actual output sequence and the one of the obtained model,

(7) *if* (small or steady state error) *then* stop *elseif* filtering of data by means of one of the procedures described in the next section and go back to (3) *endif.*

4.4.2.1 Modified Clarke's filter

In the linear case, Clarke's method amounts to determining in a progressive manner a set of filters which, when applied to the input and output sequences, lead to a decrease of the residue variance. Indeed, the perturbation that is brought by the noise can be represented in the following way (fig. 4.2): $B(z^{-1})$ and $A(z^{-1})$ represent numerator and denominator

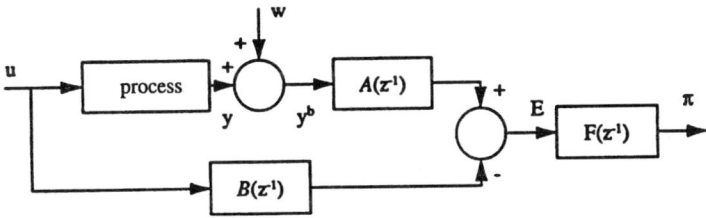

Figure 4.2: Influence of the noise.

of the transfer function respectively, and their coefficients are found again in the β and α matrices in equation (4.9), since a companion form has been chosen for the state space representation. Rather than determining the optimal filter in only one step, which means its order and its coefficients, in order to minimize the variance, it is preferred to look for a suboptimal filter of low order, usually one, and to repeat the procedure with filtered data, until the variance of the residue becomes stationary. For a first order filter, it is a matter of determining (with the unweighted least squares method) a scalar f since:

$$F(z^{-1}) = 1 + fz^{-1}. \tag{4.20}$$

Let π be the output filter and denote:

$$\begin{aligned}
\Pi &= [\pi_{m+1}, ..., \pi_N]^T \\
E &= [E_{m+1}, ..., E_N]^T \\
X_E &= [-E_m, ..., -E_{N-1}]^T
\end{aligned}$$

The scalar f is such that $\Pi^T\Pi$ is minimum, which leads to:

$$f = (X_E^T X_E)^{-1} X_E^T E. \tag{4.21}$$

At this stage, we do not dispose of the optimal filter for the residue E since the order

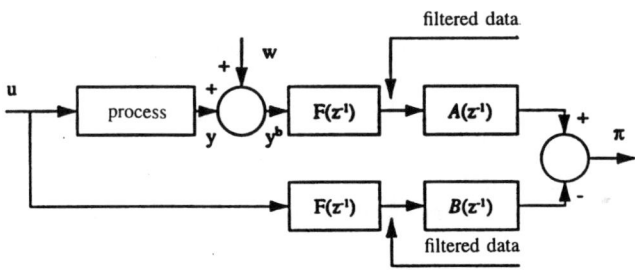

Figure 4.3: Filtering interpretation.

has been arbitrarily fixed. However instead of improving the optimization of the filter for this residue, it is applied to the input output sequences as shown in figure 4.3. It is easily checked that this scheme is equivalent to the one of figure 4.2. Then, the coefficient vector and the new residue are computed again. This means that at the ith iteration, the data that are used have been successively filtered by the Clarke's filters computed at each preceding iteration. In the present case, the problem is more difficult because the optimal filter can depend upon the values of the external parameter q. Since this one varies during the identification procedure, it amounts to determining at each iteration a varylinear filter. In practice one merely looks for a non stationary filter at each step, $F_k(z^{-1})$, without going back to the dependence upon the parameter $q(k)$. The filter $F_k(z^{-1})$ has the following form:

$$F_k(z^{-1}) = 1 + f_k z^{-1}. \tag{4.22}$$

Since the parameter q varies slowly, it is the same for f_k, which allows the use of a portion of length $(2L+1)$ of the sequence $E(j)$, the variations of the indice j being centered on k, that is $[E_{k-L}, \ldots, E_k, \ldots, E_{k+L}]$, for the computation of the optimal f_k in the least squares sense for the residue π_k. This is carried out with the relations:

$$
\begin{aligned}
E_{k-L+1} &= -E_{k-L} f_k + \pi_{k-L+1} \\
&\qquad\qquad\vdots \\
E_k &= -E_{k-1} f_k + \pi_k \\
&\qquad\qquad\vdots \\
E_{k+L} &= -E_{k+L-1} f_k + \pi_{k+L}
\end{aligned}
\tag{4.23}
$$

f_k is then given by

$$f_k = (X_{E_k}^T X_{E_k})^{-1} X_{E_k}^T \varepsilon_k \tag{4.24}$$

$$\text{where}\qquad \varepsilon_k = [E_{k-L+1}, \ldots, E_{k+L}]^T \tag{4.25}$$

$$X_{E_k} = [-E_{k-L}, \ldots, -E_{k+L-1}]^T = -\varepsilon_{k-1} \tag{4.26}$$

Data are proceeded by this time varying filter. The identification procedure is stopped when the square error between the actual process output and the one of the model, both submitted to the same inputs u and q, becomes quasi-stationary. The value of L (related to the length of the sliding memory used in the the computation of the filter) results from a compromise between the ability to take into account the variations of q (L low) and the elimination of random fluctuations by averaging (L high).

4.4.2.2 Varylinear filter

Looking again at the equation of the residue (4.15):

$$E_k = w_k - \sum_{j=0}^{\sigma_1} \sum_{i=1}^{n} a_{ij} q^j (k - i) w_{k-i} \tag{4.27}$$

with the following notations:

$$
\begin{aligned}
g_0 &= 1 \\
g_1 &= -(a_{10} + a_{11} q(k-1) + \cdots + a_{1\sigma_1} q^{\sigma_1}(k-1)) \\
& \cdot \\
& \cdot \\
g_n &= -(a_{n0} + a_{n1} q(k-n) + \cdots + a_{n\sigma_1} q^{\sigma_1}(k-n))
\end{aligned}
\tag{4.28}
$$

we obtain:
$$E_k = w_k(g_0 + g_1 z^{-1} + \ldots + g_n z^{-n}) = G(z^{-1}) w_k \tag{4.29}$$

$G(z^{-1})$ is a digital varylinear filter. $G(z^{-1})$ being known, one searches a filter $M(z^{-1})$ such that $w_k = M(z^{-1}) E_k$ with w_k equal to the white noise which is added to the output. For each instant k we have:

$$w_k = E_k - g_1 w_{k-1} - \cdots - g_n w_{k-n} \tag{4.30}$$

If initial values are chosen such that:

$$
\begin{aligned}
w_0 &= E_0 \\
&\cdots \\
w_{n-1} &= E_{n-1}
\end{aligned}
$$

equation (4.30) can be written for the instants $k = n$ up to N.

$$
\begin{aligned}
w_n &= E_n - g_1 w_{n-1} - \cdots - g_n w_0 \\
w_{n+1} &= E_{n+1} - g_1 w_n - \cdots - g_n w_1 \\
&\cdots\cdots \\
w_N &= E_N - g_1 w_{N-1} - \cdots - g_n w_{N-n}
\end{aligned}
\tag{4.31}
$$

In practice the columns of the X and Y matrices are filtered. The method can be resumed as follows:

(1) compute the first estimation of θ, i.e. $\hat{\theta} = (X^T X)^{-1} X^T Y$
(2) find the residue $E = Y - X\hat{\theta}$

(3) construct the varylinear filter and apply it to the columns of the initial matrices X and Y

(4) compute the new estimation of θ by means of filtered data with elimination of the first data that are required to initialize filtering

(5) go back to step 2

Note that in step 3, initial data are utilized and not the filtered data. As for preceding filtering, iterations are stopped when the criterion value stops increasing.

4.5 Examples

The interest for the search of a state affine model differs according to the considered process. In some cases, it simply amounts to obtaining a global model instead of a family of models usable in reduced domains. In other cases, it rather amounts to simplifying a complex model, even non-analytical, heavy and costly to exploit: a state affine model then allows a simpler representation of the process in particular working conditions. Various processes of varylinear or even nonlinear type have been modeled by state affine representations. Let us mention the electrical power production, the domains of chemistry, aeronautics or paper industry. More precisely:

- in the case of electrical power plants, where the power demand can be considered as the external parameter, these methods have been tested on steam generators of thermal or nuclear plants, with the collaboration of the Centre EDF of Clamart (Bourdon *et al.*, 1981; Dang Van Mien and Normand-Cyrot, 1984; Normand-Cyrot, 1978),

- in the case of neutralization chemical reactor, it is the volume of liquid in the tank that becomes the influential parameter (Neyran, 1987; Dadugineto, 1986),

- for a paper machine, the parameter is the outgoing speed of the sheet (Thomasset, 1987).

Hereafter two examples are presented in a succinct manner; more informations can be found in referenced publications.

4.5.1   Helicopter

This section resumes a study that was accomplished at the Laboratoire de Signaux et Systèmes of CNRS with the collaboration of the SNIAS Marignane. It concerns the representation of a helicopter in some particular flight configurations.

4.5.1.1   Presentation

The helicopter can be considered as composed of three parts (Deblon, 1986; Aksas and al., 1986):

- the aircraft, which is a rigid solid,

- the main rotor whose role is to create a force that must on one hand compensate the weight of the helicopter, and on the other hand drive the aircraft in the desired direction,

- the tail rotor which allows the control of the orientation of the aircraft.

This system has four control inputs, three of which act on the main rotor and the last on the tail rotor:

- the collective pitch rate of the main rotor controls the incidence of blades on the air, that is the lift of the rotor, or in other words the amplitude of the created force,
- the longitudinal cyclic pitch rate controls the heel, towards the front or the rear, of the rotor disk (which is defined as the part of the plane inside the circle described by the blade extremity),
- the lateral cyclic pitch rate allows the rotor disk to tilt to the side,
- the tail rotor pitch rate allows control of the transverse moment.

In order to simplify, only the aircraft motion is considered here. In this case the system is defined by eight variables:

- Six speed components (three for the translation and three for the rotation denoted u, v, w, and p, q, r respectively). These are the velocity through the mass of air in a frame attached to the aircraft.
- Two Euler's angles (trim θ, list φ) allow the change to a fixed frame.

These variables, which have a physical significance, can be chosen as state variables. Taking into account the principal rotor and tail rotor motions would require consideration of two or four more state variables.

4.5.1.2 Model

To build the model of such a system, it "suffices" to establish that the torque of the outer efforts augmented with inertial efforts is zero at the center of gravity of the aircraft. However the physical phenomena that are involved are complex and the coupling between control variables is very important: creating a simulation model requires many recordings of flight tests or wind-tunnel tests which are precise and complete the equations issued from mechanical engineering. Such a simulation program was developed at Aerospatiale and offers many possibilities to the user, but these simulations are slow and expensive. In fact, we are often faced with particular flight situations (also called equilibrium types). It is in that framework that it seems interesting to search for simplified models, each one being able to be used in the working conditions associated with the equilibrium type. Let us specify that the computations which are carried out in the above mentioned simulation program are based either on analytical equations, or on approximate equations that come from smoothing flight or wind-tunnel tests. It is not conceivable to linearize directly such a model around a working point. One solution for determining simplified models consists in executing the identification by means of input/output sequences given by the simulation program, since they are more easily obtained than in actual flight conditions. The equilibrium type which is considered here is the flight at constant altitude, with null angular velocity. Other equilibrium types concern the stabilized bank or even steady-state flights. A state affine model is searched in which the external parameter is the aircraft horizontal velocity in a terrestrial frame, which must not be confused with the velocity through the air mass in the aircraft direction. The results that are presented below concern two problems: the first deals with a SISO case, and the second focuses on the search for a multivariable model. In both cases, the state affine model is obtained through the indirect method, that is by means of regressions on the linear "tangent" models, and the input/output sequences used in the identification procedure are defined in such a way that they can be realized by a pilot.

4.5.1.3   SISO case

A model is searched linking a control input, which is the principal rotor collective pitch rate, and one output which is the aircraft velocity through the air mass. An estimation of the dimension of the linear modes was found by means of B.L. HO's algorithm, and a value of three seems reasonable. The parameter value varies by a step of 20 km/h in the range of 140–300 km/h. Indeed below 100 km/h, the aircraft behavior is quite different, which makes it difficult to search a global model valid for any speed. Various considerations allowed the determination of the degrees of polynomial expansions: 1 for the control input and up to 3 for the parameter. The results that are presented show the comparison between

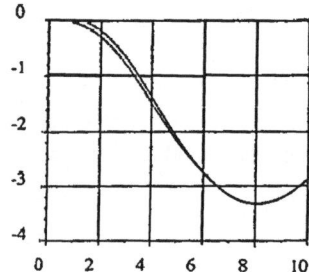 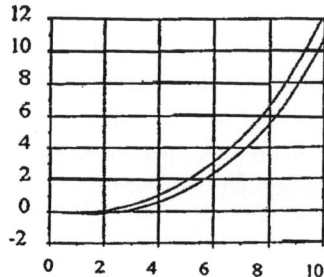

Figure 4.4: Comparison between step responses in SISO case
(input: principal collective pitch rate; output: velocity through the air mass)
(left: negative step; right: positive step; parameter 300 km/h)

the responses that are obtained by the simulation program on the one hand and by the state affine model on the other hand in two tests:

- figure 4.4a corresponds to a velocity of 300 km/h and a positive step of 0.25° as input.
- figure 4.4b corresponds to a velocity of 300 km/h and a negative step of −0.25° as input.

A good agreement of curves is verified, with an asymmetry of responses depending on whether the step is positive or negative. The simulation horizon is limited to 10 seconds, long enough allow to maintain the aircraft sufficiently close to the equilibrium in order to recover it afterwards.

4.5.1.4   Multivariable case

In this case, the state vector dimension of the linear models is equal to eight. After a few tests, it turned out that the couples of two input variables (longitudinal pitch rate or collective pitch rate) and of four output variables (translation velocity along the principal rotor axis, angular velocity around the pitching axis and both Euler's angles), can be accurately identified by means of a state affine model. Thus eight couples were identified as SISO linear models for parameter values evolving between 100 km/h and 250 km/h by a step of 50 km/h. The state affine model was built considering six monomials for the collective pitch

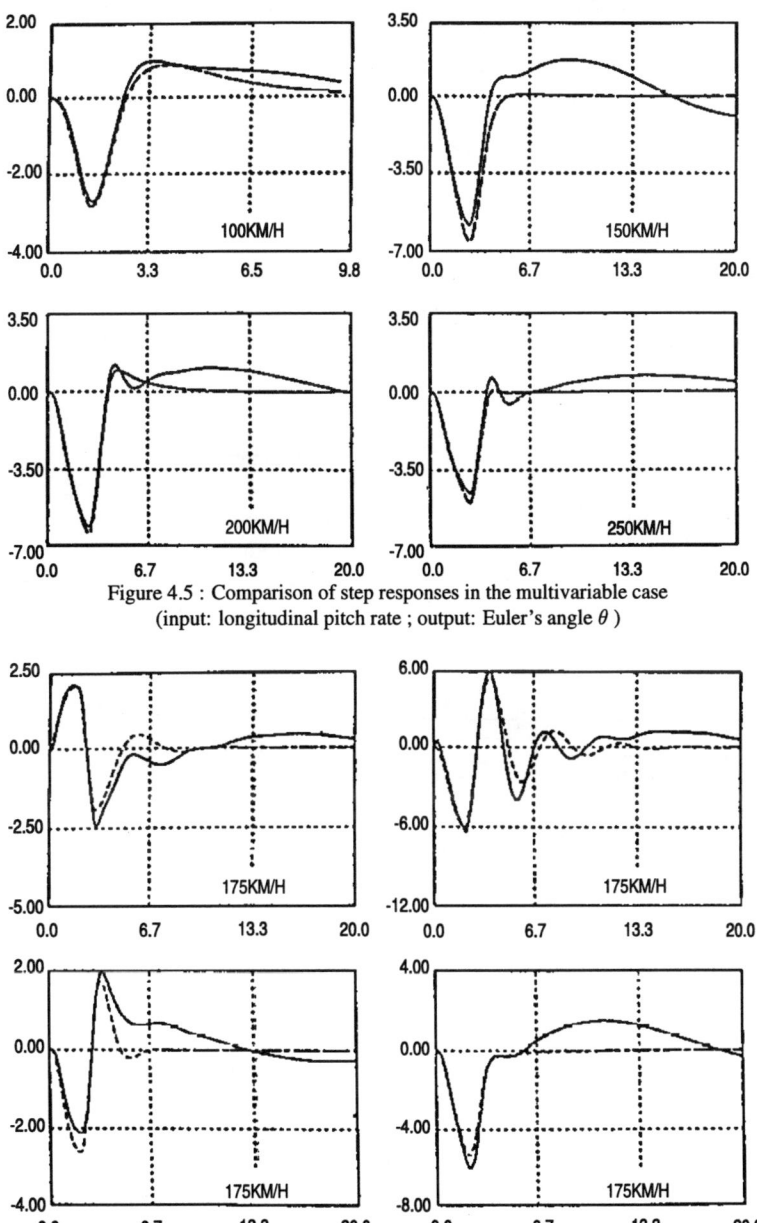

Figure 4.5 : Comparison of step responses in the multivariable case
(input: longitudinal pitch rate ; output: Euler's angle θ)

Figure 4.6 : Comparison for one parameter value which is not used for identification
(above: q and p versus principal pitch rate ; below: w and θ versus longitudinal pitch rate).

rate input and five monomials for the longitudinal pitch rate input. Figure 4.5 gives the curves corresponding to two values of the parameter, thus, allowing comparison between the responses that are given by the simulation program (continuous line), the identified linear model (long dash), and the state affine model (short dash). A good agreement of behavior is also verified. Figure 4.6 concerns a test for a parameter value that was not used for linear identification, namely 175 km/h: the outputs given by the state affine model are very close to those given by the simulation program.

4.5.2 Chemical neutralization process

The continuous neutralization reaction of a strong base (NaOH) by a strong acid (HCl) is considered. The volume in the reactor is kept constant by means of a level regulation. The equipment scheme is shown in fig. 4.7 (Neyran, 1987; Dadugineto, 1986):

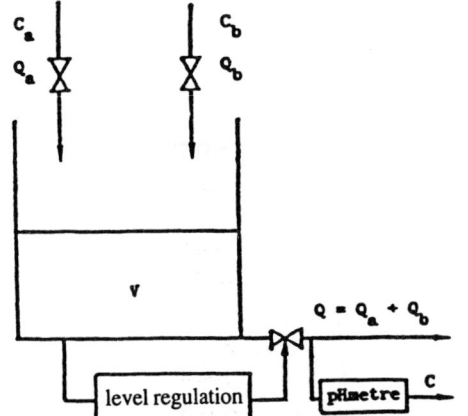

Q_a, C_a acid flow-rate and concentration as inputs;
Q_b, C_b base flow-rate and concentration as inputs;
Q output flow-rate;
$Q = Q_a + Q_b$ (level regulation)
C output concentration;
$C = C_{sa} - C_{sb}$ with:
$\quad C_{sa}$ outer acid concentration
$\quad C_{sb}$ outer base concentration
The control variable is the base flow-rate.

Figure 4.7: Neutralization process.

$$
\begin{bmatrix} x_1(k+1) \\ x_2(k+1) \\ 1 \end{bmatrix} = \left\{ \begin{bmatrix} 0 & 1 & 0 \\ 0 & .746 & 0 \\ 0 & 0 & 1 \end{bmatrix} + V(k) \begin{bmatrix} 0 & 0 & 0 \\ 0 & 4.40\,10^{-2} & 0 \\ 0 & 0 & 0 \end{bmatrix} \right.
$$

$$
+ V^2(k) \begin{bmatrix} 0 & 0 & 0 \\ 0 & -2.89\,10^{-3} & 0 \\ 0 & 0 & 0 \end{bmatrix} + V^3(k) \begin{bmatrix} 0 & 0 & 0 \\ 0 & 6.15\,10^{-5} & 0 \\ 0 & 0 & 0 \end{bmatrix}
$$

$$
+ u(k) \begin{bmatrix} 0 & 0 & 0 \\ 0 & 0 & 11.7 \\ 0 & 0 & 0 \end{bmatrix} + u(k)\,V(k) \begin{bmatrix} 0 & 0 & 0 \\ 0 & 0 & .117 \\ 0 & 0 & 0 \end{bmatrix}
$$

$$
+ u(k)\,V^2(k) \begin{bmatrix} 0 & 0 & 0 \\ 0 & 0 & -.257 \\ 0 & 0 & 0 \end{bmatrix} + u(k)\,V^3(k) \begin{bmatrix} 0 & 0 & 0 \\ 0 & 0 & 1.56\,10^{-2} \\ 0 & 0 & 0 \end{bmatrix} \left. \right\} \begin{bmatrix} x_1(k) \\ x_2(k) \\ 1 \end{bmatrix}
$$

$$
y(k) = \begin{bmatrix} 1 & 0 & 0 \end{bmatrix} \begin{bmatrix} x_1(k) \\ x_2(k) \\ 1 \end{bmatrix}
$$

This is the state space representation that is obtained by regression on 5 tangent linear models identified for $V \in \{4, 6, 8, 10, 12\}$, the liter being the unit of volume. The variable u is the base flow-rate deviation around a value Q_{b0}. A direct identification was carried out too, by decreasing the parameter V linearly with respect to time. For the same orders of polynomial expansion versus V, the state affine model that was obtained is given in Thomasset (1987). In order to compare these models, steady-state gain (fig. 4.8) and time constant (fig. 4.9) variations are given with respect to volume, for the theoretical process (continuous line), the model which is obtained by regression on several identified models (discontinuous line), and the model which is given by the direct identification (dashed line). In this example, when writing the process equations, it is shown that the steady-state gain is independent of the volume (which is the parameter), and that the time constant increases linearly versus the volume. This is approximately observed on these curves, with however a deterioration of the results for large values of the parameter. The direct method seems to lead to a better knowledge of the steady-state gain. This could be explained by the fact that in this case the identification sequence is much longer than each one used in the indirect method.

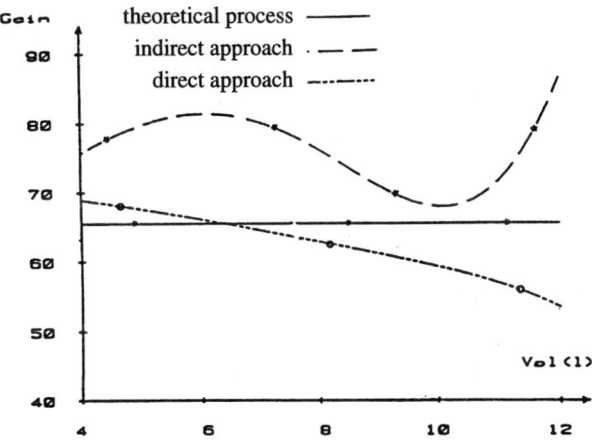

Figure 4.8: Gain variations according to the modeling.

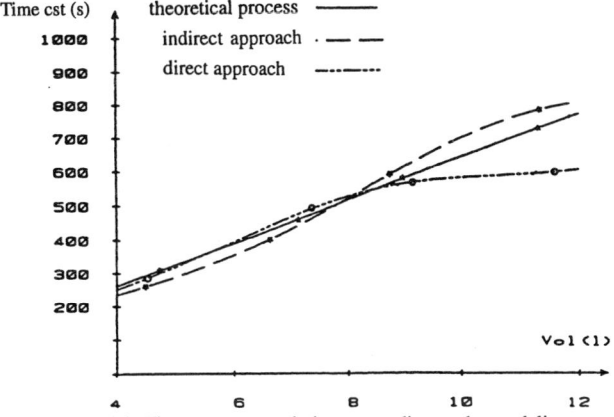

Figure 4.9: Time constant variations according to the modeling.

4.6 Presentation of AFFINE software

The aim of the AFFINE software is the calculation of the state affine model, either by means of a family of linear models, in continuous or discrete time, which are known at some working points (realization), or by means of input-output-parameter sequences (identification of a discrete representation). This AFFINE software was developed to work on an IBM PC or compatible, with an arithmetic coprocessor. It uses Matlab (version 3.2 or higher) package, with Control Toolbox, in order to utilize its capabilities concerning matrix calculations and graphic presentation of the results, together with interface modules that have been developed under Turbo Vision, for parameter acquisition by means of scrolling menus or acquisition windows, which improves user-friendliness. A first menu proposes the choice between Realization or Identification mode. Since these mode progressions are quite different, they are treated separately in this section.

4.6.1 Realization

The first operation to do is the data capture. This is carried out with the keyboard by answering the questionnaire after selecting the appropriate function. The user supplies the system models at each working point, either in a transfer function or state space representation form, in continuous or discrete time. A syntactical analyzer filters these data in real time. Once the data are collected, the user may call the calculation function which asks for the regression orders and determines the corresponding state affine model. The third step which concerns the results analysis. Several possibilities are offered to the user:

– To visualize the step or impulse responses, of the state affine and linear tangent models at each working point used for the realization, which allows the comparison.

– To visualize the obtained model response for an intermediate value of the parameter.

The user can also print or save the obtained representation. Here is an example of the windows offered:

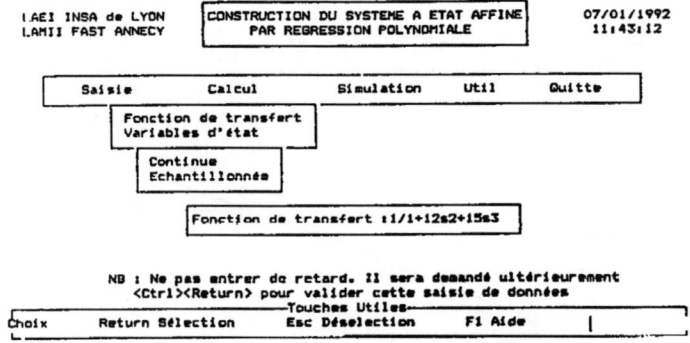

Figure 4.10: Example of a window for realization.

4.6.2   *Identification*

Here the AFFINE software needs information issued from the process, which requires a prior step for shaping these data before the phase of identification and analysis of the results.

4.6.2.1   Creation of a data file.

This stage is processed outside the AFFINE software, under the Matlab environment. Three sequences of information are supposed to be held, of the same size which may not be known, collected with a format which may differ depending on whether these data are directly collected on a process or have been generated by simulation. The Matlab load command allows the user to recover them in a suitable format. Let e, s, p be the so generated data (the name is not important, but the input, the output and the parameter must be found in that order). To make the required information complete, one must add the sampling period, let us say T. The following Matlab command $>> a = [T, 0, 0; s, e, p]$; save example a; creates a file example.mat suitable for the software which will have to find the various variables as well as the sequence length.

4.6.2.2   State affine model identification

The AFFINE software is used to carry out this stage. It must be noted that all the asked questions have default responses and that is the case for the data file which has the default name: ex1.mat. Such a file is provided with the software, which allows a quick test by simply confirming the default answers. Here also a syntactical analyzer filters the data, which allows the elimination of a certain number of errors. The first choice which is presented concerns the identification method or more precisely the filtering method which is used (see section 4.4). Then the required parameters are collected as well as the file name; for this last, the user has the possibility to list the set of files with the .mat extension in the current directory and to select the one of interest. Here is another example of the proposed windows:

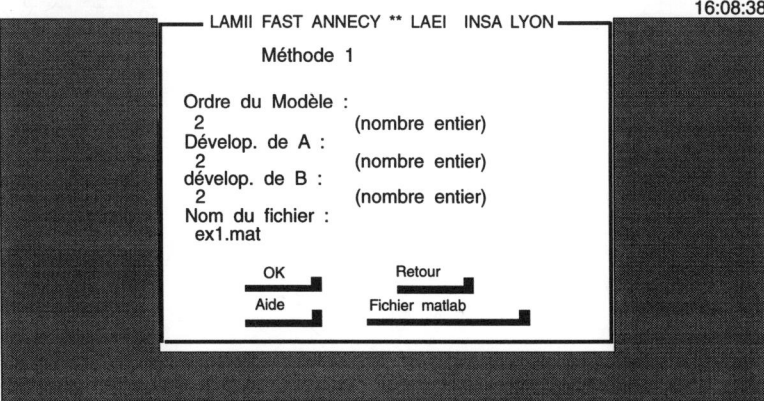

Figure 4.11: Example of a window in the Identification mode.

4.6.2.3   Analysis of results

After the identification, the user can analyze the results in two ways:
- By step response inspection. However, there is no possible comparison with the process this time. Nevertheless it is possible to follow the evolution of the step response in the course of the iterations that are required by the noise filtering process. This also permits the comparison with other models that are obtained with different orders, for example.
- By studying the main model characteristics namely its steady state gain or pole and zero evolution with respect to the external parameter. These informations may guide the user in the choice of state affine model orders. Let us note that a user who knows the Matlab language well can have access to the various data and do other processing by creating or modifying the supplied files. Figure 4.12 gives an example of the evolution of poles and zeros. It is possible to print or to save the obtained model, which may be used directly under Matlab environment.

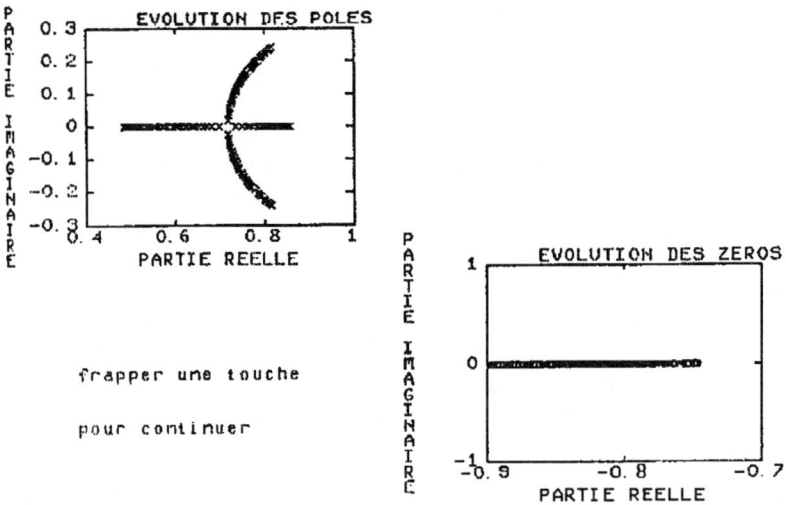

Figure 4.12: Evolution of poles and zeros.

4.7   Application to systems having an amplitude nonlinearity around each working point

Up to now we have focused on varylinear systems, which induces a particular form of the state affine representation that is searched (see equation 4.8). This section presents a utilization of the state affine representation, considered in its general form, to model a steady state asymmetry around a working point. This was developed by Bourdon (1982) for the identification of a steam generator. For an educational purpose, a system with dimension 2 is used in this discussion. Let us consider the linear system:

$$x_{k+1} = \begin{bmatrix} -a_1 & 1 \\ -a_2 & 0 \end{bmatrix} x_k + \begin{bmatrix} g_1 \\ g_2 \end{bmatrix} u_k$$

$$y_k = \begin{bmatrix} 1 & 0 \end{bmatrix} x_k$$

(4.32)

Its steady state gain is given by:

$$K = \frac{g_1 + g_2}{1 + a_1 + a_2}$$

(4.33)

and its dynamics are fixed by the polynomial roots:

$$\lambda^2 + a_1\lambda + a_2 = 0$$

By using the results in section 4.3.2, this system can be put in a bilinear or regular form, with:

$$\xi_k = \begin{bmatrix} x_k \\ 1 \end{bmatrix}$$

$$\xi_{k+1} = \left[\begin{bmatrix} -a_1 & 1 & 0 \\ -a_2 & 0 & 0 \\ 0 & 0 & 1 \end{bmatrix} + \begin{bmatrix} \alpha_{11} & \alpha_{12} & g_1 \\ \alpha_{21} & \alpha_{22} & g_2 \\ 0 & 0 & 0 \end{bmatrix} u_k \right] \xi_k$$

(4.34)

In this representation, coefficients α_{ij} are introduced in order to quit the linear framework, this case corresponding to null coefficients, and to model some nonlinearities. From null initial conditions, the state evolution, as a response to a step input u_0, is given by:

$$\xi_k = \begin{bmatrix} -a_1 + \alpha_{11}u_0 & 1 + \alpha_{12}u_0 & g_1u_0 \\ a_2 + \alpha_{21}u_0 & \alpha_{22}u_0 & g_2u_0 \\ 0 & 0 & 1 \end{bmatrix}^k \begin{bmatrix} 0 \\ 0 \\ 1 \end{bmatrix}$$

and $\quad y_k = \begin{bmatrix} 1 & 0 & 0 \end{bmatrix} \xi_k$

(4.35)

This output is identical to the one given by the linear system:

$$\tilde{x}_{k+1} = \begin{bmatrix} -a_1 + \alpha_{11}u_0 & 1 + \alpha_{12}u_0 \\ -a_2 + \alpha_{21}u_0 & \alpha_{22}u_0 \end{bmatrix} \tilde{x}_k + \begin{bmatrix} g_1 \\ g_2 \end{bmatrix} \tilde{u}_k$$

$$\tilde{y}_k = \begin{bmatrix} 1 & 0 \end{bmatrix} \tilde{x}_k$$

(4.36)

to which a unit step is applied. $\tilde{u}_k = u_0$ Let us denote this equation:

$$\tilde{X}_{k+1} = \tilde{F}\tilde{x}_k + \tilde{G}\tilde{u}_k.$$

In these conditions, the steady state gain is given by:

$$\tilde{K} = [10] \left[I - \tilde{F} \right]^{-1} \tilde{G}$$

or:

$$\tilde{K} = \frac{g_1 + g_2 + (g_2\alpha_{12} - g_1\alpha_{22})u_0}{1 + a_1 + a_2 + u_0(a_2\alpha_{12} - \alpha_{21} - \alpha_{11} - \alpha_{22}(1 + a_1)) + (\alpha_{11}\alpha_{22} - \alpha_{12}\alpha_{21})u_0^2}$$

(4.37)

and the poles are given by the roots of:

$$\lambda^2 + \lambda\left[a_1 - (\alpha_{11} + \alpha_{22})u_0\right] - (a_1 - \alpha_{11}u_0)\alpha_{22}u_0 + (1 + \alpha_{12}u_0)(a_2 - \alpha_{21}u_0) = 0$$

If only an asymmetry of steady state gain is considered, only one nonzero α_{ij} may be retained. A practical choice consists in taking

$$\alpha_{11} = \alpha_{22} = \alpha_{12} = 0.$$

One then obtains:

$$\tilde{K} = \frac{g_1 + g_2}{1 + a_1 + a_2 - \alpha_{21}u_0}.$$

To identify the state affine model around a working point, Bourdon proposes the following method:

- Collect the step responses of the process by considering two steps with the same amplitude u_0 but with opposite signs.
- Determine the average step response, which allows the identification of the linear system (a_1, a_2, g_1, g_2) .
- Determine the steady state gains K_+ and K_- corresponding to the positive and negative steps.
- Solve the system of equations:

$$K_+ = \frac{k(g_1 + g_2)}{1 + a_1 + a_2 - \alpha_{21}u_0} \tag{4.38}$$

$$K_- = \frac{k(g_1 + g_2)}{1 + a_1 + a_2 + \alpha_{21}u_0} \tag{4.39}$$

The coefficient k is introduced in order to take into account the fact that the average step response does not necessarily correspond to the one of the linear tangent model. It then results in solving the linear system with respect to k and α_{21} which has a solution if $K_+ + K_- \neq 0$. This method has been used to identify a steam generator.

4.8   Extension to MIMO systems

4.8.1   Introduction

The problem of extending state affine systems modeling techniques to MIMO varylinear processes was treated in Bourdon (1982) for the example of a steam generator (one input, four output system). The use of a set of regressions leads to the following state affine model which is searched in the same way as in the SISO case. However now the problem of the minimality of the obtained realization is questioned. It has been shown that the minimal dimension of a bilinear system is equal to the rank of the associated Hankel's matrix (Fliess, 1978). This result which concerns bilinear systems is still applicable to state affine systems; indeed, if one additional input is attached to each monomial in the input $u_1, u_2, \cdots, u_m$, the state space representation of the obtained state affine system is clearly identical to the one of bilinear discrete time systems. However, the dimension of the Hankel's matrix H (Bourdon, 1982) being usually very large, the authors prefer to calculate the rank of the observability matrix O and of the semi-accessibility matrix A (which verify $H = O^T \cdot A$). Numerically speaking, the problem is not trivial since the dimensions of these matrices increase very quickly with the number of inputs. One may use, as in Bourdon (1982), the Householder triangularization with a test of norm growth in order to obtain the rank of O (respectively of A) and thus the dimension of the minimal realization. The next sections describe a different method, which is faster, and proposed by Thomasset (1987). The process starts from an *a priori* knowledge of the rank of the searched state affine representation, and thus avoids the computation of the observability and semi-accessibility matrices. On the other hand, the use of the singular value decomposition (Lawson and Hanson, 1974) leads in a natural way to the transformation matrix and to the reduced order state affine system.

4.8.2 Recall: semi-accessibility and observability of a bilinear system

Let us consider the discrete time bilinear, also called regular, system:

$$x_{k+1} = A_0 x_k + \sum_{i=1}^{m} u_{i,k} A_i x_k \qquad (4.40)$$
$$y_k = C x_k$$

$$x_k \in R^n \; ; \; u_k \in R^m \; ; \; y_k \in R^p \; ; \; x_0 \text{ given}$$

This bilinear system is said to be semi-accessible if the set of accessible vectors spans the whole state space. Algebraically speaking (Fliess, 1978), this means that from x_0, the family of vectors:

$$A = [x_0, A_0 x_0, A_1 x_0, \ldots, (A_{j\sigma} \ldots A_{j0})^\sigma x_0, \ldots]$$
$$j_\sigma \ldots j_0 = 0, \ldots, m \qquad (4.41)$$
$$\sigma = 1, \ldots, n - 2$$

has a rank n. The observability notion is the same as for the stationary linear systems. The previously defined bilinear system is said completely observable if the matrix:

$$O = \left[C^T, ((C A_{j\sigma} \ldots A_{j0})^T)^\sigma, \ldots \right]$$
$$j_\sigma \ldots j_0 = 0, \ldots, m \qquad (4.42)$$
$$\sigma = 1, \ldots, n - 2$$

has rank n.

Remark 1

A bilinear system is reduced if it is semi-accessible and observable. Moreover, if a minimal realization of dimension μ is known for a linear invariant system, the associated bilinear system is of dimension $\mu + 1$.

4.8.3 Obtention of the reduced state affine system

Without loss of generality, for the class of systems under consideration (i.e. varylinear MIMO), we consider that the discrete realization for each value of q in the range $[q_{min}, q_{max}]$

is minimal and of dimension n (indeed, if the dimension of the minimal realizations of the linear tangent models is $\mu < n$, it suffices to replace n by μ). Considering the last remark of the previous section, the dimension of the searched reduced state affine system is $n+1$. Thus, knowing *a priori* the rank of the minimal realization, it is pointless to calculate globally the semi-accessibility and observability matrices; it is enough to build step by step each of these matrices up to the obtention of $n+1$ independent vectors.

4.8.4 Algorithm

Silverman's result (1971) concerning the minimal realization in a linear context may be transposed in the bilinear case, the problem being in both cases the calculation of the transformation matrix T such that:

If O is a matrix $m, p(m > p)$ of rank $n < p$, there exists $T(p, p)$ such that:

$$O \cdot T = m\big\{ [\underbrace{O^*}_{n} \ \underbrace{0}_{p-n}] \tag{4.43}$$

The calculation of T can be carried out by means of the singular value decomposition algorithm (Lawson and Hanson, 1974). This result allows to put the matrix O in the form:

$$O = U \begin{bmatrix} S & 0 \\ 0 & 0 \end{bmatrix} V^T. \tag{4.44}$$

U and V of dimension $m \times m$ and $p \times p$ are orthogonal. S is the diagonal matrix of the nonzero singular values, of dimension n, with:

$$s_1, s_2, \ldots, s_n \neq 0$$

Thus, if s_j is a null singular value of O, the corresponding columns u_j and v_j of matrices U and V satisfy the relation $O.v_j = s_j.u_j = 0$ for $j = n+1, p$. It follows that:

$$O.V = U \begin{bmatrix} S & 0 \\ 0 & 0 \end{bmatrix} = [O^* \ \ 0] \tag{4.45}$$

One can put: $T = V$.

4.8.5 Example

Let us consider the varylinear MIMO system given for each value of $q \in [1, 6]$, in the following transfer matrix form:

$$\begin{bmatrix} y_1(s) \\ y_2(s) \end{bmatrix} = \begin{bmatrix} \dfrac{K_1 e^{-T_1 s}}{1 + T_1 s} & \dfrac{K_2 e^{-T_2 s}}{1 + T_2 s} \\ \dfrac{K_3 e^{-T_3 s}}{1 + T_3 s} & \dfrac{K_4 e^{-T_4 s}}{1 + T_4 s} \end{bmatrix} \begin{bmatrix} u_1(s) \\ u_2(s) \end{bmatrix}$$

with $\tau_1 = 15, \tau_2 = 17, \tau_3 = 19, \tau_4 = 21, \forall q \in [1, 6]$
and:

$$q = 1 \ \Big\langle \ \begin{array}{llll} K_1 = 0,1 \ ; & K_2 = 1,1 \ ; & K_3 = 2,1 \ ; & K_4 = 3,1 \\ T_1 = 60 \ ; & T_2 = 50 \ ; & T_3 = 40 \ ; & T_4 = 30 \end{array}$$

$$q = 2 \ \Big\langle \ \begin{array}{llll} K_1 = 0,2 \ ; & K_2 = 1,2 \ ; & K_3 = 2,2 \ ; & K_4 = 3,2 \\ T_1 = 64 \ ; & T_2 = 54 \ ; & T_3 = 44 \ ; & T_4 = 34 \end{array}$$

$$q = 3 \ \Big\langle \ \begin{array}{llll} K_1 = 0,3 \ ; & K_2 = 1,3 \ ; & K_3 = 2,3 \ ; & K_4 = 3,3 \\ T_1 = 68 \ ; & T_2 = 58 \ ; & T_3 = 48 \ ; & T_4 = 38 \end{array}$$

$$q = 4 \ \Big\langle \ \begin{array}{llll} K_1 = 0,4 \ ; & K_2 = 1,4 \ ; & K_3 = 2,4 \ ; & K_4 = 3,4 \\ T_1 = 72 \ ; & T_2 = 62 \ ; & T_3 = 52 \ ; & T_4 = 42 \end{array}$$

$$q = 5 \ \Big\langle \ \begin{array}{llll} K_1 = 0,5 \ ; & K_2 = 1,5 \ ; & K_3 = 2,5 \ ; & K_4 = 3,5 \\ T_1 = 76 \ ; & T_2 = 66 \ ; & T_3 = 56 \ ; & T_4 = 46 \end{array}$$

$$q = 6 \ \Big\langle \ \begin{array}{llll} K_1 = 0,6 \ ; & K_2 = 1,6 \ ; & K_3 = 2,6 \ ; & K_4 = 3,6 \\ T_1 = 80 \ ; & T_2 = 70 \ ; & T_3 = 60 \ ; & T_4 = 50 \end{array}$$

The various time delays and time constants as well as the sampling period $(\Delta T = 15)$ are expressed with the same unit of time. The application of the realization methodology, with expansion orders equal to 2, leads to the following state affine system of dimension 13:

$$x_{k+1} = \left[\sum_{j=0}^{2} q^j(k) \cdot A_j + \sum_{j=0}^{2} u_{1,k} \cdot q^j(k) \cdot B_{1,j} + \sum_{j=0}^{2} u_{2,k} \cdot q^j(k) \cdot B_{2,j} \right] \cdot x_k \quad (4.46)$$

$$y_k = C \cdot x_k$$

The linear tangent models, at each equilibrium point parametrized by q, have a controlability matrix of rank 13 and an observability matrix of rank 8. The application of previous results shows that the state affine system is semi-accessible and that its minimal realization is of dimension $n = 9$. By means of the singular value decomposition algorithm, the observability matrix of the state affine system is built step by step until the obtention of 9 independent vectors and one finds the transformation matrix T. The state matrices of the reduced state affine system denoted $\underline{A}_j$ and $\underline{B}_{ij}$ are obtained in the following way:

$$\begin{aligned} \underline{A}_j &= \{V^T \cdot A_j \cdot V\} &\quad \text{for } j \text{ from 0 to 2} \\ \underline{B}_{ij} &= \{V^T \cdot B_{ij} \cdot V\} &\quad \text{for } i \text{ from 1 to 2 and } j \text{ from 0 to 2} \end{aligned}$$

where the operation $\{ \dots \}$ represents the removal of the first 9 lines and columns of the matrices. As for the output matrix $\underline{C}$, it is the sub-matrix $(9, 2)$ extracted of the matrix $C \cdot V$; The reduced state vector is given by the first 9 components of $V^T \cdot x_0$.

Note: the numerical values of the state matrices of the state affine system before and after reduction are found in Thomasset (1987).

4.9 Conclusion

As a conclusion, considering the current state of the art of nonlinear modeling, one can say that the method which is presented here may provide an interesting representation technique in many practical situations. Indeed, when varylinear modeling can reasonably be supposed, one then has a tool like AFFINE software at one's disposal, permitting the obtention of only one nonlinear representation in compact discrete time, over the working domain under consideration, easily exploitable for the control law synthesis, since it consists of a state model. In a more general point of view, when even in the case of small variations the process under consideration is nonlinear, the methodology that was presented can produce with a simple model, leading to faster simulations, and allowing the illustration, or even the analysis, in a more refined way of the system behavior. Considering several external parameters does not *a priori* introduce new theoretical problems (however, numerical analysis problems are still to be taken into account). For the realization, it suffices to consider polynomial regressions on several independent variables: the choice of particular values for external parameters that are used in linear tangent models is difficult. Concerning the direct identification method, the number of parameters involve an increase in the number of monomials that must be considered in the least squares problem, which leads to the handling of large dimension matrices and to the deterioration of the quality of the obtained model. Furthermore, the problem of simultaneous variations of the external parameters in order to explore the whole domain is questioned. The realization technique by means of a state affine model was extended to MIMO systems, in the case of only one parameter, possibly presenting a fixed delay and for which the rank of the minimal realization is *a priori* known. One can also mention a dual application of the realization technique to the synthesis of varylinear controllers: a family of controllers is determined for a set of configurations of the varylinear process. Then the state affine controller is built by means of regressions on this family (Dufour and Thomasset, 1984).

4.10   Bibliography

[1]   AKSAS S., DANG VAN MIEN H., DEBLON F., NORMAND-CYROT D. (1986). *Application of nonlinear identification method to helicopter flight*. Proc. IMACS IFAC Symp. Modelling and Simulation for Control of Lumped and Distributed Parameter Systems, Lille, France, pp. 291–293 .

[2]   BORNARD G., HU L.P. (1987). *Affine realizations of multimodels characterization, stability, identification*, Congrès d'Automatique, Nantes, France.

[3]   BOURDON P. (1982). *Techniques non linéaires en temps discret d'identification et de réalisation minimale pour modèles à état affine*, Doctorate thesis, Paris XI, Orsay, 142 pages.

[4]   BOURDON P., DANG VAN MIEN H., FLIESS M., NORMAND-CYROT D. (1981). *Méthodes d'identification et de réalisation non linéaires par espace d'état appliquées à des centrales nucléaires*, AFCET Congrès Automatic, Nantes, France, pp. 517–529.

[5]   CLARKE P. W. (1967). *Generalized least squares estimation of the parameters of a dynamic model*. IFAC Symposium Ident. & Syst. param. iden., Prague, Tchecoslovakia.

[6]   DADUGINETO (C. DARMET, J. DUFOUR, G. GILLES, B. NEYRAN, D. THOMASSET) (1986). *Identification de systèmes continus non linéaires. Application à deux processus pilotes*, Conférence Internationale CNRS, Paris, 1985, Nonlinear Control and Systems Theory (M. Fliess and M. Hazewinkel, eds.), Reidel, Dordrecht, pp. 597–621.

[7] DANG VAN MIEN H., NORMAND-CYROT D. (1984). *Nonlinear state-affine identification methods. Application to electrical power plants*, IFAC Symposium on Aut. Control on Power Generation, Distribution and Protection, Pretoria, pp. 449–462, *Automatica*, Vol. 20, N 2, pp. 175–188.

[8] DEBLON F. (1986). *Identification non linéaire appliquée à la mécanique du vol d'un hélicoptère.* Thesis, Paris XI, Orsay, 97 pages.

[9] DUFOUR J., THOMASSET D. (1984). *Definition and synthesis of state-affine control algorithms. Application to a varylinear system*, IEEE Conference on Computers, Systems and Signal Processing, Bangalore, India, pp. 496–499.

[10] FLIESS M. (1978). Un codage non commutatif pour certains systèmes échantillonnés non linéaires. *Info. and Cont.*, 38, pp. 264–287.

[11] FLIESS M., NORMAND-CYROT D. (1980). Vers une approche algébrique des systèmes non linéaires en temps discret, Analysis and Optimization of Systems, Versailles, *Lect. Notes Control Inform. Sciences*, N 28, Springer Verlag, Berlin, pp. 594–603.

[12] LAWSON C.L. and HANSON R.J. (1974). *Solving least squares problems*, Prentice Hall, New Jersey, 340 pages.

[13] LJUNG L. (1987). *System Identification - Theory for the user*, Prentice Hall, New Jersey, 519 pages.

[14] MONACO S., NORMAND-CYROT D. (1987). Finite Volterra Series realizations and input-output approximations of nonlinear discrete time systems, *Int J. of Cont.*, vol. 45 n 5, pp. 1771–1787.

[15] MOUILLE P. (1990). *Identification des systèmes linéaires à paramètre variable en utilisant des modèles à état affine.* DEA Automatique Industrielle, Université de Savoie, France, 68 pages.

[16] NEYRAN B. (1987). *Identification et commande en temps discret des systèmes linéaires à paramètre variable en utilisant des modèles à état affine.* Applied informatics and automatics thesis, INSA Lyon, France, 102 pages.

[17] NORMAND-CYROT D. (1978). *Utilisation de certaines familles algébriques de systèmes non linéaires à quelques problèmes de filtrage et d'identification.* Thesis Paris VII, 142 pages.

[18] SILVERMAN L. M. (1971). Realization of linear dynamical systems. *IEEE Trans on A.C.*, Vol 16, n 6, pp. 554–567.

[19] THOMASSET D. (1987). *Réalisation et commande en temps discret des systèmes continus retardés linéaires invariants et linéaires à paramètre variable.* Thesis INSA Lyon, Claude Bernard University Lyon I, 223 pages.

[20] THOMASSET D., NEYRAN B. (1985). Identification and control of nonstationary delayed systems using state-affine representation, *Int. J. of Cont.*, Vol. 42, N 3, pp. 743–758.

CHAPTER 5

Observability and observers

G. BORNARD, F. CELLE-COUENNE, G. GILLES

5.1 Introduction

The methods presented in this set of three books are based on the state concept. In this context, the actual implementation of a control law requires at each instant the knowledge of the state or of a part of it. We assume that the following variables are known:

- The system inputs that are computed from the control algorithm,

- The "outputs" measured by sensors.

In general, for several reasons (technical realizability, cost, etc.), the output vector dimension is less than the state vector dimension. Therefore, at a given time t, the state $x(t)$ cannot be algebraically deduced from the output $y(t)$ at this time. However, under some "observability" conditions which will be explained below, the state $x(t)$ can be deduced from the input-output knowledge inside a passed time interval: $u([0, t])$, $y([0, t])$.

The aim of an observer is precisely to give an "estimation" of the running value of the state with respect to the passed inputs and outputs. As the estimation has to be processed in real time, the observer usually is itself a dynamic system.

Definition 1 *We call observer (or state reconstructor) of a dynamic system:*

$$\mathcal{S} : \begin{cases} \dot{x}(t) & = & f(x(t), u(t)) \\ y(t) & = & h(x(t)) \end{cases} \tag{5.1}$$

an auxiliary dynamic system $\mathcal{O}$ whose inputs are both the input and output vectors of the system to be observed and whose output vector $\hat{x}(t)$ is the estimated state:

$$\mathcal{O} : \begin{cases} \dot{z}(t) & = & \hat{f}(z(t), u(t), y(t)) \\ \hat{x}(t) & = & \hat{h}(z(t), u(t), y(t)) \end{cases} \tag{5.2}$$

such that $\|e(t)\| = \|\hat{x}(t) - x(t)\| \longrightarrow 0$ when $t \longrightarrow \infty$.

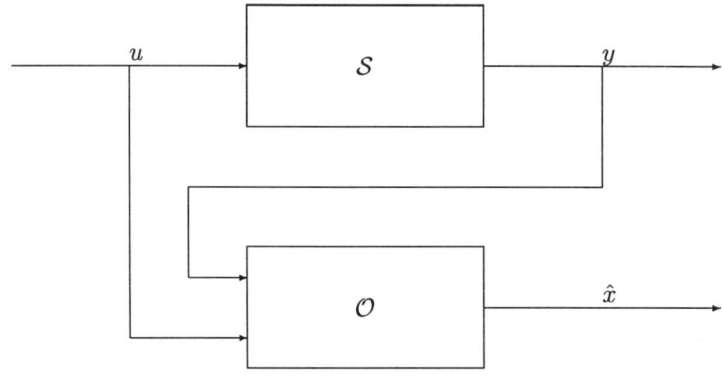

Figure 5.1: Observer.

The block diagram of such an observer is given in figure 5.1.

Remark 1

The preceding presentation of the observer function lets think that this problem is exclusively the unknown initial state problem. For the systems working under continuous conditions (industrial processes for example), the initial state concept does not have a significant meaning. However, each non-measured perturbation which disturbs the system is perceived by the observer as a "reinitialization". Then the observer has to estimate a state which depends upon this perturbation. Nevertheless, we will be able to develop all the useful tools from definition 1, on condition that the following properties are satisfied:

(i)   The dynamic characteristics of the observer must be stationary, in a sense that will be specified in section 5.3.

(ii) The estimation error must be bounded for all bounded disturbance.

(iii) The convergence speed of the observation error must be freely specified so that the observer dynamics matches the perturbations dynamics.

It is desirable for the observer to have the following property:

(iv) If the observer initialization is such that $\hat{x}(0) = x(0)$, then $\hat{x}(t) = x(t)$ for every $t > 0$.

We easily verify that an observer of the following type:

$$\mathcal{O}: \begin{cases} \dot{\hat{x}}(t) &= f(\hat{x}(t), u(t)) + k(z(t), (h(\hat{x}(t)) - y(t))) \\ \dot{z}(t) &= \hat{f}(z(t), u(t), y(t)) \end{cases} \quad \text{with } k(z(t), 0) = 0 \quad (5.3)$$

satisfies condition (iv). All the classical observers belong to this type. The extension z may not exist as it is in general the case for constant parameter linear systems.

The observers which are described in this chapter all involve the properties (i) to (iv).

Before searching an observer, we must find out whether it is possible to get a solution. The observability concept and some properties of the inputs which are applied to the system and which are presented in section 5.2 give necessary conditions for an observer synthesis.

The observability concept plays an equivalent role for the observer synthesis as the one played by identifiability for parameter estimation (chapters 3 and 4).

Section 5.3 is devoted to the observer synthesis for state affine systems and includes recalls on the linear case.

Section 5.4 deals with the case of nonlinear systems involving a strong observability property: the uniform observability, or observability for any input.

Section 5.5 is concerned with a larger class of systems and with inputs which have a local persistence property.

Section 5.6 shows how the preceding results can be extended via immersion and output injection techniques.

Section 5.7 approaches the problem of a separation principle for the nonlinear systems.

Finally, section 5.8 presents observer synthesis for a bioprocess, a mechanical system and a distillation column.

In what follows, we will consider the following systems, with $x \in \mathbb{R}^n$, $u \in \mathbb{R}^m$, $y \in \mathbb{R}^p$:

- General nonlinear systems (previously defined):

$$\begin{cases} \dot{x}(t) & = & f(x(t), u(t)) \\ y(t) & = & h(x(t)) \end{cases} \tag{5.3}$$

- Control affine systems:

$$\begin{cases} \dot{x}(t) & = & f(x(t)) + \sum_{i=1}^m u_i(t) g_i(t) \\ y(t) & = & h(x(t)) \end{cases} \tag{5.4}$$

- State affine systems:

$$\begin{cases} \dot{x}(t) & = & A(u(t))x(t) + B(u(t)) \\ y(t) & = & C(u(t))\, x(t) \end{cases} \tag{5.5}$$

- Bilinear systems:

$$\begin{cases} \dot{x}(t) & = & Ax(t) + \sum_{i=1}^m u_i(t) D_i x(t) + Bu(t) \\ y(t) & = & Cx(t) \end{cases} \tag{5.6}$$

- Time varying linear systems:

$$\begin{cases} \dot{x}(t) & = & A(t)x(t) + B(t)u(t) \\ y(t) & = & C(t)x(t) \end{cases} \tag{5.7}$$

- Constant linear systems:

$$\begin{cases} \dot{x}(t) & = & Ax(t) + Bu(t) \\ y(t) & = & Cx(t) \end{cases} \qquad (5.8)$$

Assumptions:

- The systems which are considered are analytic: all the functions which are involved in their definition are analytic (see also the appendices A1 to A3 of chapter 2, vol. 2).

- The input functions are measurable, bounded and such that the system solution is defined on $\mathbb{R}^+$.

We use the following notations:

- $X_u(t, x_0)$: Solution at time t of the system (5.1) excited by the input u initialized with x_0 at $t = 0$.

- $\Phi(t, t_0)$: Transition matrix of the free part of the system (5.7). It verifies:

$$\begin{cases} \dfrac{d}{dt}\Phi(t, t_0) & = & A(t)\Phi(t, t_0) \\ \Phi(t_0, t_0) & = & \text{Id} \quad \text{where Id is the identity matrix} \end{cases} \qquad (5.9)$$

- $\Phi_u(t, t_0)$: Transition matrix of the free part of the system (5.7) generated by the input function $u(t)$ applied to the system (5.5) –without the term $B(u(t))$.

- The norms used here are, without a contrary indication, the $\mathbb{R}^n$ Euclidian norm and the corresponding induced norm for the operators.

5.2   Observability, universality and persistence

The aim of an observer is to estimate the state, in a sense which has been previously described. This requires that the knowledge of the input and output functions upon a time interval $[0, t[$, with $t > 0$, allows any initial states pairs to be distinguished.

We can point out several related concepts which will be explained in details in the following sections:

- The *observability* of the system is related to the existence, for each initial states pair, of an input which allows such a separation.

- The *universality* of an input means the possibility, for this given input, to distinguish all the initial point pairs.

- The *regular persistence* of an input is a required property for building an observer.

5.2.1   Observability

The observability is defined from the indistinguishability concept.

Definition 2 *(indistinguishability) Considering the system 5.1, two states x_0 and x_0' are said to be indistinguishable if, for any input function u(t) and for all $t \geq 0$, the corresponding outputs $h(\mathcal{X}_u(t, x_0))$ and $h(\mathcal{X}_u(t, x_0'))$ are equal.*

Definition 3 *(observability) System (5.1) is said to be observable if it does not involve indistinguishable pairs $\{x_0, x_0'\}$ of distinct initial states.*

For the constant parameter linear systems, the observability is characterized by the well known "rank condition":

Theorem 1 *The linear system (5.8) is observable if and only if the rank of the matrix $[C \ CA \ \dots \ CA^{n-1}]$ is equal to the dimension n of the state space.*

Remark 2

Let us consider the vector space $\mathcal{O}$ of the functions of $\mathbb{R}^n$ into $\mathbb{R}^p$ spanned by the functions $Cx, CAx, \dots, CA^{n-1}x$, and the space $d\mathcal{O}$ of the (constant) differentials of each of these functions. In each point x, the evaluation for x of $d\mathcal{O}$ is precisely spanned by the columns of the matrix $[C, CA, \dots, CA^{n-1}]$.

This leads us to define, for nonlinear systems, the observation space, which will allow expression of a rank condition.

Definition 4 *(observation space) Let us consider the system (5.1). The observation space is the smallest vectorial subspace of functions of $\mathbb{R}^n$ whose values belong to the output space, which contains $h_1, \dots, h_p$, and which is closed for the Lie derivation with respect to the vector fields of the type $f_u(x) = f(x, u)$, $u \in \mathbb{R}^n$ being fixed.*

The space which is spanned by the functions $Cx, CAx, \dots, CA^{n-1}x$ is the observation space of the system (5.8).

Remark 3

One can ask if there exists an equivalence of the rank condition for the nonlinear systems. The observability concept is global: each point is distinguishable from all the others, even if they are far from each other. On the contrary, a rank condition has a local feature. We can only hope for a partial equivalence.

Let us call $d\mathcal{O}$ the space of the differentials of the elements of $\mathcal{O}$. If its evaluation $d\mathcal{O}(x)$ for x has a rank n for any x, we say that the system satisfies the rank condition for observability.

The rank condition is related to a weak observability concept. We can refer, for example, to Hermann and Krener (1977), for a synthesis of this problem, which will not be detailed here.

The case of time-varying linear systems needs a special definition:

Definition 5 *(complete uniform observability) Let us consider the time-varying linear system:*

$$\begin{cases} \dot{x}(t) &= A(t)x(t) + B(t)u(t) \\ y(t) &= C(t)x(t) \end{cases} \tag{5.7}$$

This system is said to be completely uniformly observable if there exist $T > 0$, $\alpha > 0$ and $t_0 > 0$ such that, for every $t \geq t_0$ we have:

$$\Gamma(t, t+T) = \int_t^{t+T} \Phi^T(\tau, t) C^T(\tau) C(\tau) \Phi(\tau, t) \, d\tau \geq \alpha Id \tag{5.10}$$

Remark 4

- *The matrix $\Gamma(t, t+T)$ is called the observability Grammian (at time t) of the system (5.7). If $\Gamma(t, t+T)$ is regular, the knowledge of the input and output during the time interval $[t, t+T]$ leads to the knowledge of the state in this interval. This Grammian will also be used in section 5.3.2.*

- *The complete uniform observability concept is introduced here only for time-varying linear systems.*

- *This writing is equivalent to that of Kalman (1960), when the functions $A(t)$ and $C(t)$ are bounded on $\mathbb{R}^+$, which will always be the case in the following.*

5.2.2   Universal inputs

If a linear system (5.8) is observable, for any input $u(t)$, we can reconstruct the initial state. In fact, if we consider two initial states x_0 and x_0', the quantity:

$$y'(t) - y(t) = C(\chi_u(t, x_0') - \chi_u(t, x_0)) = Ce^{At}(x_0' - x_0) \tag{5.11}$$

does not depend upon the input.

This property is not true in general for nonlinear systems. The fact that a system is observable in the sense of definition 3 leads to a necessary but non sufficient condition in order to design an observer.

As a matter of fact, some inputs u may not allow any distinct initial state pairs to be distinguished, as shown in the following example:

Example 1 Let us consider the bilinear system:

$$\begin{cases} \dot{x} = u \begin{bmatrix} 0 & 1 \\ -1 & 0 \end{bmatrix} x \;, & u \in \{0, 1\} \\ y = x_1 \end{cases} \tag{5.12}$$

It is easy to verify that this system is observable: for instance, with $u \equiv 1$, the linear system which is generated is observable. However, for $u \equiv 0$, two different values of x_2 are indistinguishable. The linear system so generated is not observable. Obviously, it is possible to design an observer working neither with $u \equiv 0$ nor with a neighboring input which may cause sensitivity problems.

Let us consider the general system (5.1). Each input u generates a time-varying homogeneous system:

$$\begin{cases} \dot{x} & = \tilde{f}(x(t), t) \quad \text{with } \tilde{f}(x(t), t) = f(x(t), u(t)) \\ y(t) & = h(x(t)) \end{cases} \tag{5.13}$$

In general, there exist singular inputs for which the homogeneous generated system is not observable. For a given system, searching these special inputs is a difficult problem, which is still largely unsolved.

Definition 6 *(universal input) An input function u is said to be universal for system (5.1) on the time interval $[0, t]$ if any distinct initial states pair $\{x_0, x_0'\}$ can be distinguished by means of the outputs on the interval $[0, t]$, the system being excited by u, that is to say if there exists $\tau \in [0, t]$ such that $h(\chi_u(\tau, x_0')) \neq h(\chi_u(\tau, x_0))$.*
An input which is universal on $\mathbb{R}^+$ is said to be universal.
A non-universal input is said to be singular.

It directly follows from this definition that:

- The homogeneous system (5.13) generated by a universal input u is observable.

- If $0 < t_1 < t_2$, then an input which is universal on $[0, t_1]$ is also universal on $[0, t_2]$.

We know how to define a universality index of the input u for state affine systems. For the system (5.5), let:

$$\Gamma_u(t, t_0) = \int_{t_0}^t \Phi_u^T(\tau, t) C^T(u(\tau)) C(u(\tau)) \Phi_u(\tau, t) \, d\tau \geq \alpha \mathrm{Id} \tag{5.14}$$

The universality index $\gamma_u(t, t_0)$ is the smallest singular value of $\Gamma_u(t, t_0)$.
For the state affine systems (5.5), the following definition is equivalent:

Definition 6' *An input u is universal on $[0, t]$ for (5.5) if $\gamma_u(t, 0) > 0$.*

For the analytic systems, the analytic input functions are dense in the set of analytic functions for the topology induced by C^∞ (Sussmann, 1979).

5.2.3 Uniformly observable systems

The universal input concept allows us to define an interesting class of systems: the uniformly observable systems.

Definition 7 *(uniformly observable system) A system where all the inputs are universal is said to be uniformly observable, or also, by abuse, observable for any input.*
If, for any $t > 0$, all the inputs are universal on $[0, t]$, the system is said to be uniformly locally observable.

Remark 5
   - *The local observability concept could also have been defined in section 5.2.1. This did not seem very useful because this concept is not different from simple observability for analytic sytems.*
   - *An observable linear system is uniformly locally observable.*

We will consider a sufficiently "regular" class of single input-single output affine systems, for which we can give a necessary and sufficient condition of uniform local observability.

Let us consider the input affine system (5.4) for which $y(t) \in \mathbb{R}$. Moreover, we assume that there exists an open, relatively compact physical domain $\Omega \in \mathbb{R}^n$ in which the state evolves and which is the interesting domain of the system.

The first step consists in writing (5.4) in a "normal" form. Let us assume that (5.4) is observable, that $u \equiv 0$ is a universal input and that the Jacobian of $\{\xi_1, \ldots, \xi_n\} = \{h, L_f h \ldots L_f^{n-1} h\}$ with respect to x has a rank n at the point x_0.

In a neighborhood of x_0, this defines a local coordinate system for which the system (5.4) can be written as:

$$\begin{cases} \dot{x}_1 &=& x_2 + \bar{\varphi}_1(x)u \\ &\vdots& \\ \dot{x}_{n-1} &=& x_n + \bar{\varphi}_{n-1}(x)u \\ \dot{x}_n &=& \tilde{\varphi}_n(x) + \bar{\varphi}_n(x)u \\ y &=& x_1 \end{cases} \tag{5.15}$$

Then, we make the following assumptions:
   - The chosen function ξ is a diffeomorphism of Ω on $\xi(\Omega)$.
   - ξ and φ can be extended from Ω to $\mathbb{R}^n$ by means of a C^∞ function.
   - The system (5.15) is complete for all the admissible input functions which are measurable and bounded and whose values belong to $\mathcal{U}$

so that (5.15) globally defines a system which coincides with the system (5.4) on Ω.

Now we can express the following result (Gauthier and Bornard, 1981; Nijmeijer, 1982; Gauthier, Hammouri and Othman, 1992):

Theorem 2 *(local uniform observability) Let us assume that system (5.4) can be expressed under the form (5.15) and satisfies the preceding assumptions. Then it is uniformly locally observable if and only if the function φ is of the following type:*

$$\begin{array}{rcl} \bar{\varphi}_1(x) &=& \bar{\varphi}_1(x_1) \\ \bar{\varphi}_2(x) &=& \bar{\varphi}_2(x_1, x_2) \\ &\vdots& \\ \bar{\varphi}_{n-1}(x) &=& \bar{\varphi}_{n-1}(x_1, \ldots, x_{n-1}) \end{array} \tag{5.16}$$

Proof:
Sufficiency: Let us consider two initial states x and $\tilde{x}$ such that $x_k = \tilde{x}_k$ for $k = 1, \ldots, i$ and $x_{i+1} \neq \tilde{x}_{i+1}$. From (5.16), one has $\bar{\varphi}_k(x) = \bar{\varphi}_k(\tilde{x})$ for $k = 1, \ldots, i$. For any u, we

obtain $\dot{x}_i - \dot{\tilde{x}}_i = x_{i+1} - \tilde{x}_{i+1}$. Thus there exists a time $t_0 > 0$ such that $x_i(t) \neq \tilde{x}_i(t)$ for every t such that $0 < t < t_0$. The iteration of this approach allows us to conclude that we obtain the same result for $\tilde{x}_1$, that is to say the output. x and $\tilde{x}$ are distinguished by any control in an arbitrary small time interval.

Necessity: Let us assume that there exists subscripts i and j, $i < j$, such that $\bar{\varphi}_i(x) = \bar{\varphi}_i(x_1, \ldots, x_i, x_j)$. Let x and $\tilde{x}$ be two initial states such that $x_k = \tilde{x}_k$ for $k = 1, \ldots, i$ and $\bar{\varphi}_i(x) \neq \bar{\varphi}_i(\tilde{x})$. Let us apply the control:

$$u(t) = -\frac{(x_{i+1}(t) - \tilde{x}_{i+1}(t))}{(\bar{\varphi}_i(x(t)) - \bar{\varphi}_i(\tilde{x}(t)))}. \tag{5.17}$$

There exists a time interval $[0, t_0[$, $t_0 > 0$ upon which $u(t)$ is defined and the system (5.15) has a solution. A simple calculation shows that $\dot{x}_k(t) = \dot{\tilde{x}}_k(t)$, $k = 1, \ldots, i$, on $[0, t_0[$. The input u thus calculated is not universal on this interval.

5.2.4   Regularly persistent inputs

Let us again consider the example 1 of section 5.2.2.

Example 1 (continued) The input $u \equiv 1$ is universal. In contrast, the input $u \equiv 0$ is singular. The function $\tilde{u}$, equal to 1 until $t_0 > 0$ and equal to 0 afterwards, is also a universal input. Let us assume that an observer exists for this system and let us apply the input $\tilde{u}$.

If a disturbance occurs on x_2 at a time $t_1 > t_0$, it does not influence the output. Therefore, the observer cannot react against this disturbance.

The universal inputs are therefore not sufficient to guarantee good properties for an observer in the presence of disturbances. This leads us to introduce a stronger concept of regular persistence, that we know how to define for state affine systems.

Definition 8 *(regularly persistent input) An input function u is said to be regularly persistent for the state affine system (5.5) if there exist $T > 0$, $\alpha > 0$ and $t_0 > 0$ such that $\gamma_u(t + T, t) \geq \alpha$ for every $t \geq t_0$.*

Remark 6

A regularly persistent input u is universal. Not only its translated function $u_\delta(t) = u(t+\delta)$ remains universal for δ arbitrarily large (persistence), but also it remains universal with a guaranteed "quality" (regularity).

Let us recall the example of the preceding section.

Example 1 (continued) The input u_1 such that:

$$u_1(t) = \begin{cases} 0 & \text{if}\quad 2kT \leq t < (2k+1)T \\ 1 & \text{if}\quad (2k+1)T \leq t < (2k+2)T \end{cases} \quad k \in \mathbb{N} \tag{5.18}$$

is regularly persistent.

This input is singular at most on uniformly bounded time intervals, which is necessary according to the definition. One can think that this property would be sufficient to guarantee the regular persistence. This is not true, as shown by the input u_2 defined by:

$$u_2(t) = \begin{cases} 0 & \text{if } 2kT \leq t < \left(2k + \dfrac{1}{k+1}\right)T \\ 1 & \text{if } \left(2k + \dfrac{1}{k+1}\right)T \leq t < (2k+2)T \end{cases} \qquad k \in \mathbb{N} \qquad (5.19)$$

which is universal and which involves a persistent property, but which is not regularly persistent.

5.3   Kalman observers for state affine systems

We have seen that the linear systems do not display the difficulties which are encountered for the synthesis of nonlinear observers. Nevertheless, they constitute an important stage from which extensions will be carried out.

5.3.1   *Observation of constant parameter linear systems*

Let us consider the linear system (5.8). Designed according to the principle presented in section 5.1, the observer has the following state equation:

$$\dot{\hat{x}} = A\hat{x} + Bu + K(y - Cx). \qquad (5.20)$$

Denoting $e(t) = \hat{x}(t) - x(t)$, the error equation satisfies the homogeneous equation:

$$\dot{e} = (A - KC)e. \qquad (5.21)$$

We know that, if the system (the pair A, C) is observable, then we can choose the observer gain matrix K such that $(A - KC)$ has a negative real part spectrum which implies the convergence towards zero of $e(t)$ when $t \to \infty$.
One can even arbitrarily set this spectrum; see Luenberger (1966). The specification of the gain K can be realized in different ways: poles placement, constant solution of a Riccati equation, etc.

5.3.2   *Time varying linear systems, Kalman observers*

The problem of state estimation of linear dynamic systems disturbed by noises has been solved by Kalman and Bucy (1960).

In this presentation, we give the deterministic version of the Kalman filter.

Let us consider the time varying linear system, assumed completely uniformly observable:

$$\begin{cases} \dot{x}(t) & = & A(t)x(t) + b(t) \\ y(t) & = & C(t)x(t) \end{cases} \qquad (5.22)$$

where $b(t) = B(t)u(t)$. The functions $A(t), C(t)$ are assumed to be bounded.

Theorem 3 *Let us consider the completely uniformly observable system (5.22):*

$$
\begin{cases}
\dot{\hat{x}}(t) &= A(t)\hat{x}(t) - S^{-1}(t)C^T(t)Q(C(t)\hat{x}(t) - y(t)) + b(t) \\
\dot{S}(t) &= -\theta S(t) - A(t)^T S(t) - S(t)A(t) + C^T(t)QC(t) - S(t)RS(t) \quad (5.23) \\
S(0) &= S_0 \quad (S_0 \text{ positive definite})
\end{cases}
$$

If one of the two following conditions is satisfied:

   (i) the pair $(A(t), R(t))$ is completely uniformly controlable,
   (ii) $\theta > 0$,

then the system (5.23) is an observer for the system (5.22), whose estimation error norm is bounded by a decreasing exponential function. Moreover, the choice of θ allows an arbitrary decreasing velocity to be reached.

Remark 7

   - *The introduction of the "exponential forgetting factor" θ also allows to work with $R = 0$. In all the following (including sections 5.4 and 5.5), we make this assumption which simplifies the computations. Nevertheless, all the results that are presented are true with the complete Kalman formalism.*
   - *The property of complete uniform controllability is the dual property of the complete uniform observability. Since $R = 0$ in the following, this concept will not be described here.*

The interpretation of the Kalman observer in terms of output error minimization in the least square sense is given by the following theorem (Couenne, 1988):

Theorem 4 *The estimated state given by the previous observer (with $R = 0$) is a solution of the following minimization problem:*

$$
\hat{x}(t) \quad : \quad \min_{\hat{x}} J(\hat{x}) = \|\Phi(0,t)\hat{x} - \hat{x}_0\|_{S_0}^2 + \int_0^t \|\Phi(\tau,t)\hat{x} - y(\tau)\|_Q^2 \, d\tau \quad (5.24)
$$

where $\|x\|_Q^2$ means $x^T Q x$.

5.3.3   Observation of state affine systems

This section will be devoted to systems of the form (5.5). Let us remark that these systems involve, in general, singular inputs. We show that, under conditions on the input, the theory recalled in section 5.3.2 can be applied.

The system (5.5) being given, for any admissible input corresponds the linear system of type (5.8) in which $A(t) = A(u(t))$, $b(t) = B(u(t))$, $C(t) = C(u(t))$. We assume that $A(u), B(u), C(u)$ are uniformly bounded on the admissible input domain. The definitions 5 and 8 immediately lead to:

Theorem 5 *Any regularly persistent input applied to the state affine system (5.5) generates a time varying linear system (5.22) which is completely uniformly observable.*

The preceding property allows the observer design issued from the Kalman theory to be extended to the state affine systems (Bornard, Couenne and Celle, 1987; Couenne, 1988).

Theorem 6 *(observer for state affine systems) Let us consider the state affine system (5.5) in which $A(u)$, $B(u)$, $C(u)$ are uniformly bounded upon the admissible input domain. Then, for any regularly persistent input of (5.5), the system:*

$$\begin{cases} \dot{\hat{x}} = A(u)\hat{x} - S^{-1}C^T(u)Q(C(u)\hat{x} - y) + B(u) \\ \dot{S} = -\theta S - A(u)^T S - SA(u) + C^T(u)QC(u) \end{cases} \quad \text{with } \theta > 0 \qquad (5.25)$$

is an observer for the systems (5.5), whose estimation error norm is bounded by a decreasing exponential. Moreover, an arbitrary decreasing velocity can be chosen.

Remark 8

  – *As a particular case, the observer (5.25) works for bilinear systems.*
  – *When $R = 0$ in the observers (5.24) or (5.25), the criterion (5.24) minimization is processed in an n- dimensional space, n being the state space dimension.*

5.4   High gain observers for uniformly locally observable systems

5.4.1   Introduction

We have just seen how a Kalman observer design can be used for state affine systems excited by regularly persistent inputs. Taking into account the flexibility of this approach, it is natural to try and extend it to a larger class of systems.

The "high gains" type methods allow us to reach this goal for certain classes of systems. The principle which is applied is the following:

  • The system is decomposed into a linear or state affine part and a nonlinear part.
  • The observer synthesis is realized on the linear part basis.
  • Under structural conditions, the use of a sufficient "gain" allows us to take into account an arbitrary nonlinearity for the second part.

After a presentation of the basic assumptions and of the formalism which is used, three situations of increasing complexity are considered:

  • single output uniformly locally observable systems
  • multi output uniformly locally observable systems
  • systems excited by "locally regular" inputs.

The first case allows us to develop calculations in a simple situation and presents the advantage to lead to a necessary and sufficient condition. The second one is a generalization of the preceding one. It leads to a sufficient condition and shows some unsolved difficulties related to the absence of observability canonical form for the multi-output nonlinear systems.

5.4.2 System and assumptions

Let us consider the input affine system (5.4), in which $u(t)$ belongs to the set $\mathcal{U} \subset \mathbb{R}^m$ of the input admissible values. Moreover, we assume that there exists a "physical domain" (open, bounded) $\Omega \subset \mathbb{R}^n$ in which the state evolves, which is the domain of interest of the system.

As for section 5.2.3, the first step consists in writing (5.4) in a "normal" form. Let us assume that (5.4) is observable and that $u \equiv 0$ is a universal input. The Jacobian matrix of $\left\{ h_1, L_f h_1, \ldots, L_f^{n-1} h_p \right\}$ with respect to x has a rank n almost everywhere on $\mathbb{R}^n$. In a neighborhood of a regular point, we can select a full rank subset:

$$\{\xi_1 \ldots \xi_n\} = \left\{ h_1, L_f h_1 \ldots L_f^{\eta_1} h_1, \ldots, h_p, \ldots, L_f^{\eta_p} h_p \ldots \right\} \qquad (5.26)$$

This defines a "pyramidal" local coordinate system in which the system (5.4) can be locally written as:

$$\begin{aligned} \dot{x} &= Ax + \tilde{\varphi}(x) + \bar{\varphi}(x)u \\ y &= Cx \end{aligned} \qquad (5.27)$$

where:

$$A = \begin{bmatrix} A_1 & & \\ & \ddots & \\ & & A_p \end{bmatrix}, \qquad A_k = \begin{bmatrix} 0 & 1 & & 0 \\ \vdots & & \ddots & \\ 0 & \ldots & 0 & 1 \\ 0 & \ldots & 0 & 0 \end{bmatrix},$$

$$C = \begin{bmatrix} C_1 & & \\ & \ddots & \\ & & C_p \end{bmatrix}, \qquad C_k = [1\ 0 \ldots 0], \qquad (5.28)$$

$$\tilde{\varphi}(x) = \begin{bmatrix} \tilde{\varphi}_1(x) \\ \vdots \\ \tilde{\varphi}(x)_p \end{bmatrix}, \qquad \tilde{\varphi}_k(x) = \begin{bmatrix} 0 \\ \vdots \\ 0 \\ \check{\varphi}_k(x) \end{bmatrix}$$

The size of each block k is η_k, $k = 1, \ldots, p$, with $\sum_{k=1}^{p} \eta_k = n$, and $\mu_1 = 1$, $\mu_k = \mu_{k-1} + \eta_{k-1}$, $k = 2, \ldots, p$.

As a matter of fact, the linearity with respect to u is not used in the next part and the systems which will be considered are in the following form:

$$\begin{aligned} \dot{x} &= Ax + \varphi(x, u) \\ y &= Cx \end{aligned} \qquad (5.29)$$

We then use the following assumptions:
- The function ξ which is chosen is a diffeomorphism of Ω on $\xi(\Omega)$,
- ξ and φ can be extended from Ω to $\mathbb{R}^n$ through a function C^∞,
- The system (5.29) is complete for all the admissible input functions (measurable and bounded) whose values belong to $\mathcal{U}$,

so that (5.29) globally defines a system which coincides with the system (5.4) on Ω.

5.4.3 Single output systems

In a first step, we consider the single output system (5.29). The following theorem is given by Gauthier, Hammouri and Othman, (1992).

In this section, the subscript k is removed from the notation.

Theorem 7 *Let us consider the system (5.29) with $p = 1$, in which:*

   (i) the function φ is globally Lipschitzian with respect to x, uniformly with respect to
   u.

Let K be an $n \times 1$ matrix such that the matrix $(A - KC)$ has all negative real part eigenvalues, and $\Lambda(T) =$
$$\begin{bmatrix} T & & & \\ & T^2 & & \\ & & \ddots & \\ & & & T^n \end{bmatrix}$$

Let us assume that:

   (ii) $\dfrac{\partial \varphi_i}{\partial x_j} \equiv 0$ for $i = 1, \ldots, n-1$, $j = i+1, \ldots, n$.

Then the system (5.29) is uniformly locally observable and there exists T_0 such that, for any T verifying $0 < T \leq T_0$, the system:

$$\dot{\hat{x}} = A\hat{x} + \varphi(\hat{x}, u) + \Lambda^{-1}(T)K(y - C\hat{x}) \tag{5.30}$$

is an observer for the system (5.29). Moreover, the observation error norm is bounded by an exponential whose decreasing velocity can be chosen arbitrarily large.

Remark 9

- *The matrix K can be chosen as the solution of a Riccati algebraic equation or as a solution of a Lyapunov equation with an exponential forgetting factor.*
- *The condition (ii) of the theorem is also necessary in order for the system (5.29) to be uniformly locally observable.*
- *The design of the observer is realized in the starting coordinate system. It implies the calculation of the Jacobian matrix $\dfrac{\partial \xi}{\partial x}$ but not the effective computation of φ in the new coordinate system.*

Proof (sketch)
The following points summarize the proof and will allow us to understand the meaning of the conditions introduced in the multi-output-case. From the notation used, we easily verify that:

$$\begin{aligned} \bar{A} &= \Lambda(T)A\Lambda^{-1}(T) &= T^{-1}A, \\ \bar{C} &= C\Lambda^{-1}(T) &= T^{-1}C. \end{aligned} \tag{5.31}$$

Let $\bar{x} = \Lambda x, \tilde{x} = \Lambda \hat{x}, \bar{\varepsilon} = \tilde{x} - \bar{x}, \psi(\bar{\varepsilon}) = \varphi(\tilde{x}, u) - \varphi(\bar{x}, u)$, the dependence of ψ with respect to x and u being omitted in the notation. Then:

$$\begin{aligned}
\dot{\bar{x}} &= T^{-1} A \bar{x} + \Lambda(T) \varphi(\Lambda^{-1}(T) \bar{x}, u) \\
y &= T^{-1} C \bar{x} \\
\dot{\tilde{x}} &= T^{-1} A \tilde{x} + \Lambda(T) \varphi(\Lambda^{-1}(T) \tilde{x}, u) + T^{-1} KC(\bar{x} - \tilde{x}) \\
\dot{\bar{\varepsilon}} &= T^{-1}(A - KC)\bar{\varepsilon} + \Lambda(T) \psi(\Lambda^{-1}(T) \bar{\varepsilon})
\end{aligned} \tag{5.32}$$

Since $(A - KC)$ is stable, there exists a positive definite solution of the Lyapunov equation:

$$(A - KC)^T S + S(A - KC) = -I \tag{5.33}$$

Let $\bar{v} = \bar{\varepsilon}^T S \bar{\varepsilon}$. We obtain:

$$\begin{aligned}
\dot{\bar{v}} &= 2\bar{\varepsilon}^T S \dot{\bar{\varepsilon}} \\
&= -T^{-1}\|\bar{\varepsilon}\|^2 + 2\bar{\varepsilon}^T S \Lambda(T) \psi(\Lambda^{-1}(T) \bar{\varepsilon}) \\
&\leq \frac{-T^{-1}}{\lambda_S} \bar{v} + 2\lambda_S \|\bar{\varepsilon}\| \alpha_\varphi \sum_{i=1}^{n} \sum_{j=1}^{n} \chi_{ij} T^{i-j} \|\bar{\varepsilon}\| \\
&\leq \left(\frac{-T^{-1}}{\lambda_S} + 2\frac{\lambda_S}{\lambda_I} \alpha_\varphi \sum_{i=1}^{n} \sum_{j=1}^{n} \chi_{ij} T^{i-j} \right) \bar{v}
\end{aligned} \tag{5.34}$$

where:

- α_φ is the Lipschitz constant of φ (condition i)),
- λ_I, λ_S are the minimum and maximum eigenvalues of S,
- $\chi_{ij} = \begin{cases} 0 & \text{if } \frac{\partial \varphi_i}{\partial x_j} \equiv 0, \\ 1 & \text{in the counter case.} \end{cases}$

From assumption (ii) of the theorem, $\chi_{ij} \equiv 0$ for $i < j$. Thus $\chi_{ij} T^{i-j} < 1$ for $T < 1$, $i = 1, \ldots, n, j = 1, \ldots, n$. This implies that:

$$\dot{\bar{v}} \leq \left(\frac{-T^{-1}}{\lambda_S} + 2\frac{\lambda_S}{\lambda_I} \alpha_\varphi n^2 \right) \bar{v} \tag{5.35}$$

Then, for every $\gamma > 0$, there exists T, $0 < T < 1$, such that $\dot{\bar{v}} \leq -\gamma \bar{v}^2$. This ends the proof of the theorem.

5.4.4 Multi-output systems

Denoting $\sigma_i, i = 1, \ldots, n$ the successive powers of T in Λ, the main points of the preceding approach can be summarized as follows:

- The relationships (5.31) come from the fact that the σ_i are equidistant ($\sigma_{i+1} - \sigma_i$ does not depend on i),
- The σ_i satisfy $\frac{\partial \varphi_i}{\partial x_j}. \equiv 0$ for $\sigma_i < \sigma_j$.

Let us now consider the system (5.29) without any restriction on p, that is to say with several outputs. In order to get an equivalent of theorem 7 for such systems, the first point must be satisfied inside each subsystem, while the second point remains the same. With the notations of section 5.4.2, we can express the following theorem:

Theorem 8 *Let us consider the system (5.29) in which:*

(i) *The function φ is globally Lipschitzian with respect to x, uniformly with respect to u.*

Let $K = \begin{bmatrix} K_1 & & \\ & \ddots & \\ & & K_p \end{bmatrix}$ be a matrix of adequate dimension such that, for each k block, the matrix $(A_k - K_k C_k)$ has all negative real part eigenvalues.

Let us assume that there exist two integer sets $\sigma = \{\sigma_1 \ldots \sigma_n\}$ and $\delta = \{\delta_1 \ldots \delta_p\}$, with $\delta_i > 0, \; ; i = 1 \ldots p$, such that:

(ii) $\sigma_{\mu_k + l} = \sigma_{\mu_k + l - 1} + \delta_k, \quad k = 1 \ldots p, \; l = 1 \ldots \eta_k - 1,$

(iii) $\partial \varphi_i / \partial x_j \neq 0 \Rightarrow \sigma_i \geq \sigma_j, \quad i = 1 \ldots n, \; j = 1 \ldots n, \; j \neq \mu_k, k = 1 \ldots p.$

Then the system (5.29) is uniformly locally observable, and there exists $T_0 > 0$ such that, for any $T, 0 < T < T_0$, the following system:

$$\dot{\hat{x}} = A\hat{x} + \varphi(\hat{x}, u) + \Lambda^{-1}(T, \delta) K (y - C\hat{x}) \tag{5.36}$$

with:

$$\hat{x}_{\mu_k} = y_k,$$
$$\hat{x}_j = \hat{x}_j, j \neq \mu_k,$$
$$\Lambda(T, \delta) = \begin{bmatrix} T^{\delta_1} \Delta(T^{\delta_1}) & & \\ & \ddots & \\ & & T^{\delta_p} \Delta(T^{\delta_p}) \end{bmatrix}, \quad \Delta_k(T) = \begin{bmatrix} 1 & & & \\ & T & & \\ & & \ddots & \\ & & & T^{\eta_k - 1} \end{bmatrix} \tag{5.37}$$

is an observer for the system (5.29). Moreover, the observation error norm is bounded by an exponential whose decreasing speed can be chosen arbitrarily large.

The proof of theorem 8, given in Bornard and Hammouri (1991), follows the same approach as that of theorem 7. Moreover, the reader interested by the specificity of the multi-output case can refer to the proof of theorem 9, which is a generalization of the previous one.

5.4.5 Remarks and example

- In the described observer, the conditions assigned to φ are rigid with respect to the structure ($\partial \varphi_i / \partial x_j \equiv 0$ for certain pairs (i, j)), and on the contrary more flexible with respect to the type of the function φ (only the Lipschitz constant is involved).
- The "high gain" terminology is used because a sufficiently small T value (T^{-1} large enough) allows us to "hide" the nonlinear part effect in the observer design.

- Inside each block, the structure conditions are a copy of the local uniform observability of a single input system.
- For the multi-output systems, they define couplings in φ for which the system remains locally uniformly observable, and an observer of type (5.36) can be designed. They also define the relative orders of importance of the gains corresponding to each subsystem.
- The test of conditions (ii) and (iii) leads to a search for the indexes σ_i. The conditions (ii) and (iii) are invariant under the translations on σ, since only the differences $\sigma_i - \sigma_j$ play a part. Moreover, from the condition (ii), the choice of the σ_i and δ_k reduces to that of $\delta_1 \ldots \delta_p$ and $\sigma_{\mu_1} \ldots \sigma_{\mu_p}$. It is easy to see that the set of relevant values to be tested is finite. So, for a system given in the form (5.29), the test of conditions (ii) and (iii) leads to a finite enumeration.

Figure 5.2 shows the admissible structures of $\dfrac{\partial \varphi}{\partial x}$ for a system with two outputs and with $\eta_1 = 3, \eta_2 = 2$. The σ_i are indicated. The investigation of these combinations and of those which are not shown on the figure provides a good understanding of the mechanism of conditions (ii) and (iii).

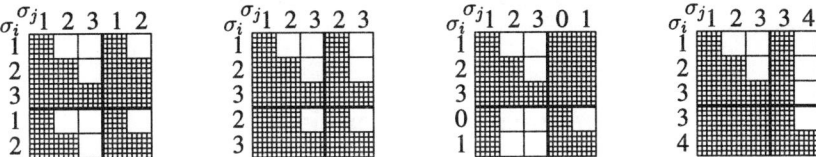

Figure 5.2: Admissible structures of $\dfrac{\partial \varphi}{\partial x}$ for a system with two outputs.

- The structure conditions depend upon the coordinates in which the system equations are written. This clearly appears in example 2.

Example 2 Let us consider the following system with two outputs:

$$\dot{x} = \begin{bmatrix} 0 & 1 & 0 \\ 2.5 & 8.75 & -.25 \\ .5 & .25 & -.75 \end{bmatrix} x + u \begin{bmatrix} -.5 & .75 & -.25 \\ 12 & 8.5 & 14.5 \\ 1 & 0 & 4 \end{bmatrix} x \tag{5.38}$$

$$y = \begin{bmatrix} 1 & 0 & 0 \\ 0 & 0 & 1 \end{bmatrix} x$$

which leads to:

$$\varphi(x, u) = \begin{bmatrix} 0 & 0 & 0 \\ 2.5 & 8.75 & -.25 \\ .5 & .25 & -.75 \end{bmatrix} x + u \begin{bmatrix} -.5 & .75 & -.25 \\ 12 & 8.5 & 14.5 \\ 1 & 0 & 4 \end{bmatrix} x \tag{5.39}$$

Under this coordinate system, which corresponds to $(h_1, L_f h_1, h_2)$, the system does not statisfy the theorem 8 conditions.

Now, by choosing the coordinate system defined by $(h_1, h_2, L_f h_2)$, we get:

$$
\dot{x} = \begin{bmatrix} 2 & 3 & 4 \\ 0 & 0 & 1 \\ 4 & 5 & 6 \end{bmatrix} x + u \begin{bmatrix} 1 & 2 & 3 \\ 1 & 4 & 0 \\ 5 & 6 & 7 \end{bmatrix} x
$$

$$
y = \begin{bmatrix} 1 & 0 & 0 \\ 0 & 1 & 0 \end{bmatrix} x
$$
(5.40)

Written in this form, the system satisfies the conditions (with $\sigma = \{2, 1, 2\}$).
From this, we deduce that the system (5.38) is locally uniformly observable, and, at the same time, we verify that the theorem 8 conditions are only sufficient with respect to the local uniform observability.

- The type of dependence upon the coordinates that has been pointed out is related to a finite combinatory, which is not tedious. However it is not the only one. Any coordinate transformation in the output space – which does not modify the observability – can also be realized before writing the system equations in the form (5.29). The question of knowing how to find a suitable base transformation through which the theorem conditions would be satisfied is open. This is due to the fact that we do not know how to build a canonical form for a non-autonomous multi-output nonlinear system.

5.5 Structured nonlinear perturbation of state affine systems

5.5.1 Class of systems taken into account

In the previous section, the observer has been designed by using only a fixed linear dynamic part. In the same way as the Kalman approach has been extended to state affine systems, the previous approach will be extended to systems which can be decomposed into a state affine part and a nonlinear part which will involve structural constraints.

We now consider a system in the following form:

$$
\begin{aligned}
\dot{x} &= A(u)x + \varphi(x, u) \\
y &= C(u)x
\end{aligned}
$$
(5.41)

with $x \in \mathbb{R}^n$, $u \in \mathbb{R}^m$, $y \in \mathbb{R}^p$,

$$
A(u) = \begin{bmatrix} A_1(u) & & \\ & \ddots & \\ & & A_p(u) \end{bmatrix}, \quad A_k(u) = \begin{bmatrix} 0 & a_1^k(u) & & 0 \\ \vdots & & \ddots & \\ 0 & \cdots & 0 & a_{\eta_k-1}^k(u) \\ 0 & \cdots & 0 & 0 \end{bmatrix}, \quad (5.42)
$$

$$
C(u) = \begin{bmatrix} C_1(u) & & \\ & \ddots & \\ & & C_p(u) \end{bmatrix}, \quad C_k(u) = [c_k(u) \, 0 \ldots 0]. \quad (5.43)
$$

η_k and μ_k, $k = 1, \ldots, p$ are defined as previously.

At present, we do not know if a given nonlinear system can be equivalent to (5.41) by a coordinate transformation. Once again, this is due to the non-existence of a canonical form. We will see that the results presented below, could, nevertheless, be applied to the observation of physical processes (distillation).

5.5.2 Locally regular inputs

A and C being functions of the input u, assumptions on the function u will have to be made in order to guarantee that the system is observable "enough". The regular persistence concept, which concerns the input behavior for $t \to \infty$, has been introduced in section 5.2.4. Because of the use of high gain techniques, we will define the local regularity concept, which, on the contrary, concerns small time intervals. The system excited by such an input will be, in some sense, "as observable" as a linear system.

This concept is defined with respect to the state affine part of the system (5.41):

$$\begin{aligned} \dot{x} &= A(u)x \\ y &= C(u)x \end{aligned} \tag{5.44}$$

Let Φ_k be the transition matrix of the k-th subsystem of (5.44) and let us consider the observability Grammians $\Gamma_k(t, T)$ of each of the k-th subsystems of (5.44):

$$\Gamma_k(t, T) = \int_{t-T}^{t} \Phi_k^T(\tau, t) C_k^T(u(\tau)) C_k(u(\tau)) \Phi_k(\tau, t) \, d\tau \tag{5.45}$$

where the u-dependency has been dropped in the expressions of Φ_k and Γ_k.

Definition 9 *An input function u is said to be locally regular if there exists $\alpha_\Gamma > 0$, $T_0 > 0$ such that:*

$$\Gamma_k(t, T) > \alpha_\Gamma T \Delta_k^2(T), \quad \forall T \leq T_0, \ \forall t \geq T, \ k = 1 \ldots p \tag{5.46}$$

where: $\Delta_k(T) = \begin{bmatrix} 1 & & & \\ & T & & \\ & & \ddots & \\ & & & T^{\eta_k - 1} \end{bmatrix}$

In order to justify this definition, let us consider the linear part of the system (5.29). The Grammian can be written in the following form:

$$\Gamma(t, T) = T \begin{bmatrix} 1 & \dfrac{T}{2} & \dfrac{T^2}{6} & \cdots \\ \dfrac{T}{2} & \dfrac{T^2}{6} & \dfrac{T^3}{8} & \\ \dfrac{T^2}{6} & \dfrac{T^3}{8} & \dfrac{T^4}{20} & \\ \vdots & & & \ddots \end{bmatrix} \tag{5.47}$$

Then:

- The lower boundedness which is involved in the local regularity definition is – up to a constant matrix – the lowest that can be proposed.
- For an observable linear system, any input is locally regular.

5.5.3   Observer synthesis

Theorem 9 *Let us consider the system (5.41) in which:*

(i)   *The function φ is globally Lipschitzian with respect to x, uniformly with respect to u, and A, C are bounded on $\mathcal{U}$..*

Let us assume that there exist two integer sets $\sigma = \{\sigma_1 \ldots \sigma_n\}$ and $\delta = \{\delta_1 \ldots \delta_p\}$, with $\delta_i > 0$, $i = 1 \ldots p$, such that:

(ii) $\sigma_{\mu_k+l} = \sigma_{\mu_k+l-1} + \delta_k, \quad k = 1, \ldots, p, \ l = 1, \ldots \eta_k - 1$

(i) $\partial \varphi_i / \partial x_j \neq 0 \Rightarrow \sigma_i \geq \sigma_j, \quad i, j = 1, \ldots, n, \ j \neq \mu_k, k = 1, \ldots, p$

(iv)   *the input u is locally regular for the system (5.44).*

Then there exist $T > 0$, $\beta > 0$, $\alpha > 0$, such that the following system:

$$\dot{\hat{x}} = A(u)\hat{x} + \varphi(\hat{x}, u) + S^{-1}C^T(u)(y - C\hat{x}) \tag{5.48}$$

$$\dot{S} = -\Theta S - A^T(u)S - SA(u) + C^T(u)C(u) \tag{5.49}$$

$$\text{with:} \quad \begin{cases} S(0) = \begin{bmatrix} S_1^0 & & \\ & \ddots & \\ & & S_p^0 \end{bmatrix}, \quad \Theta = \begin{bmatrix} \theta_1 I & & \\ & \ddots & \\ & & \theta_p I \end{bmatrix}, \\[2em] S_k^0 = \alpha \Delta_k^2(T^{\delta_k}), \quad \theta_k = \beta T^{-\delta_k}, \quad k = 1, \ldots, p \\[0.5em] \hat{x}_{\mu_k} = y_k, \ \hat{x}_j = \hat{x}_j, \ j = \mu_k + 1, \ldots, \mu_k + \eta_k - 1, \quad k = 1, \ldots, p. \end{cases}$$

is an observer for the system (5.41). Moreover, the observation error norm is bounded by an exponential whose decreasing velocity can be chosen arbitrarily large.

Before giving a sketch of the theorem proof, a technical lemma is required for estimating upper and lower bounds of S. The proof of this lemma will be omitted.

First, let us notice that the solution of the Lyapunov equation (5.49) is block-diagonal if S^0 is chosen in this form. In this case, each block is driven by:

$$\dot{S}_k = -\theta_k S_k - A_k^T(u)S_k - S_k A_k(u) + C_k^T(u)C_k(u), \quad k = 1 \ldots p \tag{5.50}$$

Let us denote by $S_k(t, \theta_k, S_k^0)$ the solution of the equation (5.50) with $S_k(0) = S_k^0$.

Lemma 1 *Let us assume that A_k and C_k are bounded on $\mathcal{U}$, and that the input u is locally regular for (5.44), with α_Γ and T_0 as local regularity constants. Then, the following properties are true:*

(i) $\Delta_k^{-1}(T)S_k(t, \theta_k, S_k^0)\Delta_k^{-1}(T) \geq e^{-\theta_k T}\alpha_\Gamma T \ I, \ \forall S_k^0 > 0, \ \forall T \leq T_0, \ \forall t \geq T, \ k = 1, \ldots, p$

(ii)   *there exist $\beta > 0$, $\alpha > 0$, $\alpha_S > 0$, such that:*

$$\|\Delta_k^{-1}(T)S_k(t, \beta T^{-1}, \alpha \Delta_k^2(T))\Delta_k^{-1}(T)\| \leq \alpha_S T, \ \forall T \leq T_0, \ \forall t \geq T, \ k = 1, \ldots, p \tag{5.51}$$

Proof of theorem 9:

With the following notations:

$$e = \hat{x} - x, \quad \begin{cases} \breve{e}_{\mu_k} &= 0, \\ \breve{e}_j &= e_j, \ j = 1, \ldots, n, \ j \neq \mu_k, \end{cases} \quad k = 1, \ldots, p$$

$$\varepsilon_k = \begin{bmatrix} e_{\mu_k} \\ \vdots \\ e_{\mu_k + \eta_k - 1} \end{bmatrix}, \quad \breve{\varepsilon}_k = \begin{bmatrix} \breve{e}_{\mu_k} \\ \vdots \\ \breve{e}_{\mu_k + \eta_k - 1} \end{bmatrix}, \quad v_k = \varepsilon_k^T S_k \varepsilon_k, \quad k = 1, \ldots, p \tag{5.52}$$

$$\psi(\breve{\varepsilon}) = \varphi(x + \breve{\varepsilon}, u) - \varphi(x, u),$$

we have:

$$\begin{aligned}
\dot{e} &= A(u)e + \psi(\breve{\varepsilon}) - S_k^{-1} C_k^T(u) C_k(u) e \\
\dot{\varepsilon}_k &= A_k(u)\varepsilon_k + \psi_k(\breve{\varepsilon}) - S_k^{-1} C_k^T(u) C_k(u)\varepsilon_k \\
\dot{v}_k &= 2\varepsilon_k^T S_k \dot{\varepsilon}_k + \varepsilon_k^T \dot{S}_k \varepsilon_k \\
&= 2\varepsilon_k^T S_k A_k(u)\varepsilon_k + 2\varepsilon_k^T S_k \psi_k(\breve{\varepsilon}) - 2\varepsilon_k^T C_k^T(u) C_k(u)\varepsilon_k \\
&\quad -\theta_k \varepsilon_k^T S_k \varepsilon_k - 2\varepsilon_k^T S_k A_k(u)\varepsilon_k + \varepsilon_k^T C_k^T(u) C_k(u)\varepsilon_k \\
&\leq -\theta_k v_k + 2\varepsilon_k^T S_k \psi_k(\breve{\varepsilon})
\end{aligned} \tag{5.53}$$

Let $\bar{v}_k = T^{2\sigma_{\mu_k}} v_k$, $\bar{v} = \sum\limits_{k=1}^{p} \bar{v}_k$. Then:

$$\begin{aligned}
\dot{\bar{v}}_k &\leq -\theta_k \bar{v}_k + 2T^{2\sigma_{\mu_k}} \varepsilon_k^T S_k \psi_k(\breve{\varepsilon}) \\
&\leq -\tilde{q}_k + \hat{q}_k
\end{aligned} \tag{5.54}$$

We will upper bound those two terms in order to compare them.

Let us define $\Lambda = \begin{bmatrix} \Lambda_1 & & \\ & \ddots & \\ & & \Lambda_p \end{bmatrix} = \begin{bmatrix} T^{\sigma_1} & & \\ & \ddots & \\ & & T^{\sigma_n} \end{bmatrix}$, $\begin{aligned} \bar{\varepsilon}_k &= \Lambda_k \varepsilon_k, \ k = 1 \ldots p, \\ \bar{e} &= \Lambda e, \end{aligned}$.

By applying lemma 1 with $S_k^0 = \alpha \Delta_k^2(T^{\delta_k})$, and using the assumptions of the theorem, we get:

$$\begin{aligned}
\tilde{q}_k &\geq \alpha_\Gamma \beta e^{-\beta} \|\bar{\varepsilon}_k\|^2 \\
\tilde{q} &= \sum_{i=1}^{p} \tilde{q}_k \geq \tilde{\alpha} \|\breve{e}\|^2 \\
\hat{q}_k &\leq 2\alpha_S T^{\delta_I} \alpha_\varphi n \eta_k \|\bar{e}\|^2 \\
\hat{q} &= \sum_{i=1}^{p} \hat{q}_k \leq \hat{\alpha} T^{\delta_I} \|\bar{e}\|^2
\end{aligned} \tag{5.55}$$

and finally:

$$\alpha_\Gamma e^{-\beta} T^{\delta_S} \|\bar{e}\|^2 \leq \bar{v} \leq \bar{v}_0 e^{-\alpha T^{-\delta_I} t} \tag{5.56}$$

Let us remark that, in the theorem, the initial condition $S(0)$ of S has been explicitly fixed, in order to simplify the approach. In fact, any positive definite initial condition would be convenient in the statement.

5.5.4   Remarks

- The general remarks presented for theorem 8 remain valid.
- The structural assumptions for A can be weakened in the following sense: suppose that the condition $A(i,j) \equiv 0$ for $i \geq j$ is removed from the definition of $A(u)$, while $A(i,j) \equiv 0$ for $i < j - 1$ subsists. Then the same result is still valid. This lower triangular part can be set in φ, without changing the condition (iii).
- If the matrices A and C of the system (5.41) are constant and make an observable pair, theorem 9 reduces itself into theorem 8, in which a special choice of matrix K is made by using the steady-state solution of the Lyapunov equation.
- In the observer design, the Lyapunov equation (5.49) has been used in order to simplify the calculations. Nevertheless, theorem 9 can be presented with a Riccati equation instead of (5.49):

$$\dot{S} = -\Theta S - SRS - A^T(u)S - SA(u) + C^T(u)C(u) \qquad (5.57)$$

where R is positive semi-definite. In (5.57), θ can be set to zero if R is positive definite. This has been proved in Deza (1991), for a distillation column, in addition, by replacing A by $(A + \partial\varphi/\partial u)$, then giving to the observer the behavior of an extended Kalman observer around a steady-state point. This special approach (choice of the coordinate system) for designing an extended Kalman observer allows us to obtain a stability result.

5.6   Immersion and output injection

Throughout this chapter, we have been concerned with special classes of nonlinear systems. The application range of the Luenberger and Kalman observers has been extended to:

- The state affine systems, owing to a quality analysis of the inputs.
- The systems for which high gain techniques still allow us to realize the design under specified structural conditions.

Another way of extending the class of systems for which we know how to realize observer synthesis is to find a transformation which – locally or globally – "sends" a given nonlinear system (5.1) into another one, whose dimension is possibly larger, for which the already presented synthesis methods can be applied.

The immersion approach consists in finding a differentiable application which immerses the initial system state space into the arrival system state space by preserving the observability property. This approach is fundamentally interesting because the immersion keeps the observability property. If the immersion is a diffeomorphism, we call it a system equivalence.

The output injection is also an interesting transformation because it does not modify the observer synthesis problem.

These transformations can be compared, in principle, to the ones that are used for control: the transformations by means of diffeomorphism and feedback. The dimension increase related to the immersion plays the same part as the extension associated with dynamic feedback.

5.6.1 Immersion

The main results on immersion are those of Fliess and Kupka (1983) and Levine and Marino (1986). They give necessary and sufficient conditions of immersion into a bilinear system or a finite dimensional system. We will only present the result for bilinear systems.

First, let us precise the immersion concept:

Definition 10 (*immersion of a system into another one*) *Let us consider the systems described by:*

$$\begin{cases} \dot{x} &= f(x,u) \\ y &= h(x) \end{cases} \quad x \in \mathbb{R}^n, \ u \in \mathbb{R}^m, \ y \in \mathbb{R}^p \quad (5.1)$$

and:

$$\begin{cases} \dot{z} &= \bar{f}(z,u) \\ w &= \bar{h}(z) \end{cases} \quad z \in \mathbb{R}^{\bar{n}}, \ u \in \mathbb{R}^m, \ w \in \mathbb{R}^p \quad (5.58)$$

Let us call $x(t, x_0, u)$ and $z(t, z_0, u)$ the solutions at time t of the systems (5.1) and (5.58) respectively initialized by x_0 and z_0.
Let τ be a local immersion of a neighborhood of $x^* \in \mathbb{R}^n$ into $\mathbb{R}^{\bar{n}}$ (of the same differentiability degree as the systems being considered). τ is said to be a local immersion of system (5.1) into the system (5.58) near x^* if there exists a neighborhood X^* of x^* such that, for any $x_0 \in X^*$ and any admissible input function u, the following commutation properties are satisfied for any time t for which the solution of (5.1) is defined:

$$z(t, \tau(x_0), u) = \tau(x(t, x_0, u)) \quad (5.59)$$

$$\bar{h}(z(t, \tau(x_0), u)) = h(x(t, x_0, u)) \quad (5.60)$$

Then we say that (5.1) can be immersed into (5.58).
If $X^* = \mathbb{R}^n$ in this definition, the immersion is global.

Remark 10

If both systems are observable, the relationship (5.60) implies the relationship (5.59).

Theorem 10 (*Fliess-Kupka*) *The analytic system (5.1) affine with respect to the input is immersible into a bilinear system if and only if the observation space has a finite dimension.*

This result leads to an observer design for such a system. Let us assume that the system (5.1) is observable and satisfies the conditions of theorem 10. It can be immersed into a bilinear system:

$$\begin{cases} \dot{x}(t) &= Ax(t) + \sum_{i=1}^{m} u_i(t) D_i x(t) + Bu(t) \\ y(t) &= Cx(t) \end{cases} \quad (5.6)$$

Then, we are in one of the two following situations:

- The system (5.6) is observable. Then we can design an observer for this system by using the theory of section 5.3. Let us consider the projection $\tilde{z}$ of the estimated state $\hat{z}$ on the image submanifold $\mathcal{T} = \tau(\mathbb{R}^n)$ of the immersion and $\hat{x} = \tau^{-1}(\tilde{z})$. Because of the observability of the systems taken into account, $\hat{x}$ is well defined. Since (5.6) is observable, $\hat{z}(t)$ tends to $\tau(x(t))$ and it tends to $\mathcal{T}$ when $t \to \infty$. Then $\hat{x} \to x(t)$ when $t \to \infty$.

- If not, the system (5.6) involves a non observable vector subspace $\mathcal{N}$ and an observable representation in $\mathbb{R}^n/\mathcal{N}$, for which an observer can be designed. The observability of (5.1) implies that $\pi_{\mathcal{N}} \circ \tau$ is injective ($\pi_{\mathcal{N}}$ is the canonical projection on $\mathbb{R}^n/\mathcal{N}$).The observable case demonstration remains valid when replacing τ^{-1} by $\tau^{-1} \circ \pi_{\mathcal{N}}^{-1}$.

To this condition, the theorem of Levine and Marino (1986) adds another condition for the functions f, g and h of the system (5.4) which means that the system must be linear.

5.6.2 Output injection

The linearization by means of output injection, developed for the autonomous systems by Krener and Isidori (1983) and Krener and Respondek (1985), consists in transforming by means of a coordinate change, a system (5.1) into an observable linear system modulo an output injection term, of the following type:

$$\begin{cases} \dot{z} &= Az + B(u,y) \\ y &= Cz \end{cases} \tag{5.61}$$

Remark 11

If the pair (A,C) is observable, then the system (5.61) has no singular inputs, that is to say it is observable for any input. We can design an observer of the type:

$$\dot{\hat{z}} = A\hat{z} + B(u,y) + K(y - C\hat{z}) \tag{5.62}$$

where K is such that $A - KC$ is stable.

The case of autonomous systems simply introduces the approach.

Theorem 11 *(Krener-Isidori) The autonomous system:*

$$\begin{cases} \dot{x} &= f(x) \\ y &= h(x) \end{cases} \tag{5.63}$$

is locally equivalent (through a diffeomorphism) close to a state x_0 to an autonomous linear system modulo an output injection:

$$\begin{cases} \dot{z} &= Az + b(y) \\ y &= Cz \end{cases} \tag{5.64}$$

if and only if $f(x_0) = 0$, $h(x_0) = 0$ and the vector field $g(x)$ defined by:

$$L_g L_f^k(h) \equiv \begin{cases} 0 & \text{if} \quad 0 \le k < n-1 \\ 1 & \text{if} \quad k = n-1 \end{cases} \tag{5.65}$$

satisfies the relationship:

$$[g, \mathrm{ad}_f^k(g)] \equiv 0 \quad for \quad k = 1, \dots, 2n - 1 \tag{5.66}$$

with $\mathrm{ad}_f^0(g) = g$, $\mathrm{ad}_f^k(g) = [g, \mathrm{ad}_f^{k-1}(g)]$.

The equivalence to a state affine system modulo of a diffeomorphism and an output injection is a generalization of the preceding problem: it consists in transforming a nonlinear system (5.1), observable in the rank sense, into a system of the following type:

$$\begin{aligned} \dot{z} &= A(u)z + B(u, y) \\ y &= Cz \end{aligned} \tag{5.67}$$

A necessary and sufficient condition has been given by Hammouri and Gauthier, (1988) (bilinearization), 1989 (affine equivalence).

Other results on the immersion into special classes of observable state affine systems via the output injection can be found in the works of Hammouri, Celle and De Leon (1990, 1991).

5.7 Separation principle

Let us assume that we can use a state feedback control law which stabilizes the system and that an observer has been designed. In the practical realization, we want to replace, in the control law computation, the unknown real state by the estimated state (see figure 5.3). The main question is to know if the stability of the system controled in such a way is preserved.

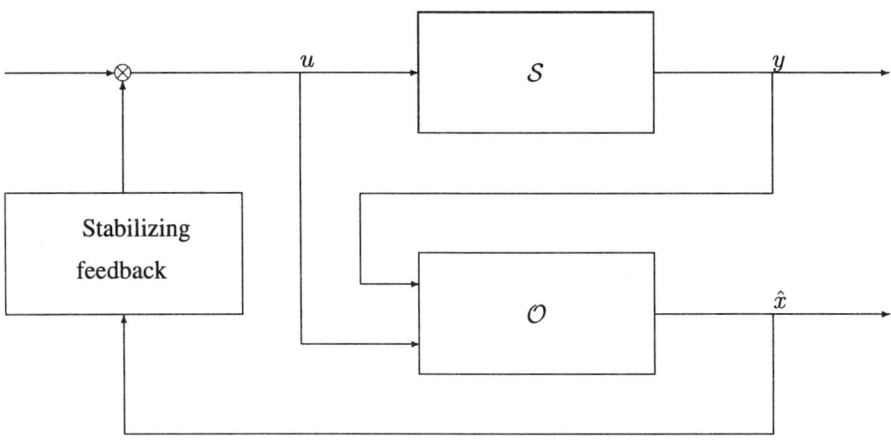

Figure 5.3: Feedback on the estimated state.

This problem, which is on the border of control and observation, is delicate. In general, we have to find a compromise between the observer convergence velocity and the stabilization speed.

It is known that, for a linear system, the spectrum of the system controled by means of a feedback from the estimated state is split into two parts: the observer spectrum and the spectrum of the feedback control system without observer, both spectra being unchanged. The stability of the feedback control system with observer thus depends only on the observer stability and on the stability of the feedback control system without observer.

This separation principle, which makes the design much easier is, in general, not applicable in the nonlinear world. However, it has been proved that some classes of nonlinear systems verify a weakened separation principle (see Gauthier, Hammouri and Kupka, 1991).

Thus, for the dissipative drift bilinear systems, a stabilizing control law computed from a Lyapunov function associated with the drift remains stable when it is looped on an observer whose type is described in section 5.3.

H. Hammouri extended this result to the nonlinear systems which are observable for any input.

Let us consider the system (5.29) controled by a feedback law $u_0(x)$ belonging to the class C^∞ and assumed to be globally asymptotically stable. We can prove (under some technical assumptions) that the system controled by the same feedback law, computed from the estimated state, calculated from an observer derived from theorem 7, is globally asymptotically stable.

5.8   Applications

5.8.1   *Bioreactor*

The example presented in this section has been developed by S. Othman at LAGEP (Othman, 1992). The system which is studied is a single input system and it is uniformly locally observable.

5.8.1.1   Process, model

The system is a bioreactor working in continuous conditions, whose scheme and notations are given in figure 5.4. In the following, we use the simplest model (representation of the biomass by a global concentration, the same for the substrate, no saturation of the growth rate, etc.). We only want to show an example of observer design.

The mass balance leads to the equations:

$$\begin{cases} V\dot{X} = \mu V X - F X \\ V\dot{S} = -\dfrac{1}{K}\mu V X + F(S_e - S) \end{cases} \qquad (5.68)$$

For the growth rate, we use one of the classical models developed by Contois (1959):

$$\mu(X, S) = \frac{a_1 S}{a_2 X + S} \tag{5.69}$$

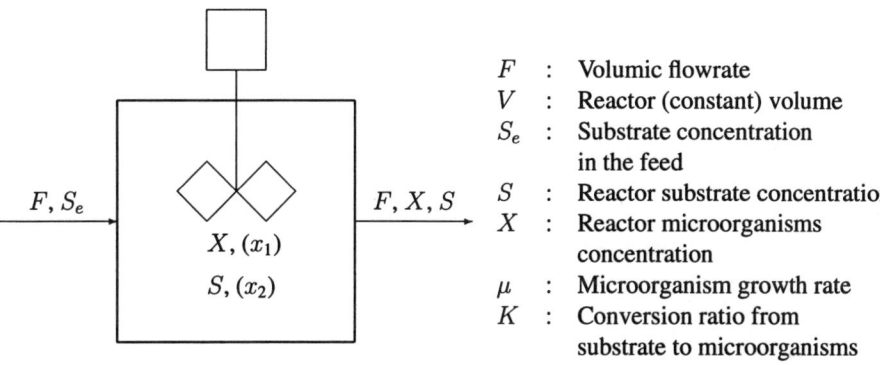

F	:	Volumic flowrate
V	:	Reactor (constant) volume
S_e	:	Substrate concentration in the feed
S	:	Reactor substrate concentration
X	:	Reactor microorganisms concentration
μ	:	Microorganism growth rate
K	:	Conversion ratio from substrate to microorganisms

Figure 5.4: Continuous bioreactor.

We assume that the feeding substrate concentration is constant. Setting $u = \frac{F}{V}$, $x_1 = X$, $x_2 = S$, $a_3 = \frac{1}{K}$, $a_4 = S_e$, and putting the μ expression in the balance equations, we obtain:

$$\begin{cases} \dot{x}_1 &= \dfrac{a_1 x_1 x_2}{a_2 x_1 + x_2} - u x_1 \\[2mm] \dot{x}_2 &= -\dfrac{a_3 a_1 x_1 x_2}{a_2 x_1 + x_2} + u(a_4 - x_2) \\[2mm] y &= x_1 \end{cases} \tag{5.70}$$

Here, the chosen measured output is the biomass. The following result can be applied in the same way with the substrate as an output.

5.8.1.2 Physical domain, coordinate transformation

We first notice that the domain

$$\mathcal{D} = \{x \mid x_1 > 0, \ x_2 > 0 \ a_4 - a_3 x_1 - x_2 > 0\} \tag{5.71}$$

is invariant for (5.70) (the same applies to its adherence $\bar{\mathcal{D}}$), if u and a_4 are positive, which is physically natural. However, in order to construct the coordinate transformation which is necessary for the observer design, it will be useful if $\bar{\mathcal{D}}$ does not contain zero. Thus, we will define an invariant domain somewhat more restricted, and we will verify that this limitation is not inconvenient for the use of the model.

Let α_2, u_s be such that $0 < \alpha_2 < a_4$, $0 < u_s < a_1$, and α_1, u_i, be given by:

$$\begin{aligned} \alpha_1 &= \frac{(a_1 - u_s)\alpha_2}{a_2 u_s} \\[2mm] u_i &= \frac{a_3 a_1 \alpha_2}{a_2(a_4 - \alpha_2)} \end{aligned} \tag{5.72}$$

Substituting (5.72) into (5.70), it comes that $\{x|\ \alpha_1 < x_1, \alpha_2 < x_2\}$ and its adherence is invariant for the system (5.70), for all the input functions whose values belong to the set $\mathcal{U}$ defined by $\mathcal{U} = \{u|\ u_i < u < u_s\}$.

Let us denote respectively by $\Omega(\alpha_2, u_s)$ the set:

$$\Omega(\alpha_2, u_s) = \mathcal{D} \cap \{x|\ \alpha_1 < x_1,\ \alpha_2 < x_2\}, \tag{5.73}$$

and $\bar{\Omega}(\alpha_2, u_s)$ its adherence. $\bar{\Omega}(\alpha_2, u_s)$ is compact.

The condition $u_s < a_1$ is not too restrictive: As a matter of fact, for $u_s > a_1$ we have $x_1 \to 0$ when $t \to \infty$, which does not correspond to a standard operation of the process. The lower bounds α_1, α_2 and u_i can be taken arbitrarily small. The restriction of the operating domain to $\Omega(\alpha_2, u_s)$ is not a serious drawback. In the following, the dependence of Ω with respect to α_2, u_s will be omitted in the notation.

The coordinate change $z = \Phi(x)$ which sets the system into its normal form 5.4 of section 5.4 is given by:

$$\Phi(x) = \begin{bmatrix} \Phi_1(x) & = & x_1 \\ \Phi_2(x) & = & \dfrac{a_1 x_1 x_2}{a_2 x_1 + x_2} \end{bmatrix} \tag{5.74}$$

The function Φ is a diffeomorphism of the first open quadrant onto itself. Ω and $\bar{\Omega}$ are included in this domain.

In the new coordinates, the state equation of the system is given by:

$$\begin{bmatrix} \dot{z}_1 \\ \dot{z}_2 \end{bmatrix} = \begin{bmatrix} z_2 \\ \varphi(z) \end{bmatrix} + \begin{bmatrix} g_1(z1) \\ g_2(z_1, z_2) \end{bmatrix} u \tag{5.75}$$

where:

$$g_1(z_1) = -z_1$$

$$g_2(z_1, z_2) = \frac{a_4 z_2^2}{a_1 a_2 z_1^2} - \frac{2 a_4 z_2}{a_2 z_1} + \frac{a_1 a_4}{a_2} - z_2$$

$$\varphi(z_1, z_2) = \left(\frac{1}{a_1} - \frac{a_3}{a_1 a_2} \right) \frac{z_2^3}{z_1^2} + \frac{2 a_3 z_2^2}{a_2 z_1} - \frac{a_1 a_3 z_2}{a_2} \tag{5.76}$$

In order to reach an adequate form for the construction of the observer, we must:

- extend the diffeomorphism $\Phi|_{\bar{D}}$ where $\bar{D}$ is compact to a diffeomorphism of $\mathbb{R}^2$ onto itself, Lipschitzian and with a Lipschitzian inverse,
- extend the functions g and φ which are Lipschitzian on $\mathbb{R}^2$.

This could be done in C^∞ without too much difficulty. In practice, a continuous connexion is sufficient.

5.8.1.3   Observer design

The computation of the matrix K of the observer is processed by means of a stationary Lyapunov equation:

$$0 = -S - A^T S - S A + C^T C$$
$$K = \Lambda^{-1}(T) S^{-1} C^T \tag{5.77}$$

where S is symmetric positive definite. In dimension 2, the solution is given by:

$$S = \begin{bmatrix} 1 & -1 \\ 1 & 2 \end{bmatrix}, \quad S^{-1} = \begin{bmatrix} 2 & 1 \\ 1 & 1 \end{bmatrix}, \quad K = \begin{bmatrix} 2T^{-1} \\ T^{-2} \end{bmatrix} \tag{5.78}$$

Coming back to the initial coordinates leads to:

$$\dot{\hat{x}}_1 = \frac{a_1 \hat{x}_1 \hat{x}_2}{a_2 \hat{x}_1 + \hat{x}_2} - u\hat{x}_1 + 2T^{-1}(y - \hat{x}_1)$$

$$\dot{\hat{x}}_2 = -\frac{a_3 a_1 \hat{x}_1 \hat{x}_2}{a_2 \hat{x}_1 + \hat{x}_2} + u(a_4 - \hat{x}_2) \tag{5.79}$$

$$+ \left(T^{-2} \left(\frac{\hat{x}_2^2}{a_1 a_2 \hat{x}_1^2} + \frac{2\hat{x}_2}{a_1 \hat{x}_1} + \frac{a_2}{a_1} \right) - 2T^{-1} \frac{\hat{x}_2^2}{a_2 \hat{x}_1^2} \right) (y - \hat{x}_1)$$

5.8.1.4 Simulation experiments

The simulation has been carried out with the following parameter and control values:

$$V = 1, a_1 = 1, a_2 = 1, S_e = 0.1, T = 1, u = \begin{cases} 0.08 & \text{for } 0 < t \le 10 \\ 0.02 & \text{for } 10 < t \le 20 \\ 0.08 & \text{for } 20 < t \end{cases}$$

Figure 5.5 shows the noisy output x_1 and the estimated states $\hat{x}_1$ and $\hat{x}_2$ as compared to the "actual states" (simulated).

5.8.2 Overhead crane

The example presented in this section has been carried out by T. Chorot at LAGEP (Chorot, 1991).

5.8.2.1 System and model

The overhead crane described in figure 5.6 is a mechanical system example whose state equations are naturally in a suitable form for an observer study. The inputs are the horizontal motion force F_r on the carriage and the lifting force F_l applied to the load. The positions r, l, α are assumed to be measured.

Using the Lagrange approach, the following equations are obtained:

$$L = T - V$$

$$\frac{d}{dt} \left(\frac{\partial L}{\partial \dot{q}} \right) - \frac{\partial L}{\partial q} = F \tag{5.80}$$

With the notations of figure 5.6 and $q = (r, l, \alpha)^T$, we have:

$$T(q, \dot{q}) = \frac{1}{2} \begin{bmatrix} \dot{r} & \dot{l} & \dot{\alpha} \end{bmatrix} \begin{bmatrix} M + N & M\sin(\alpha) & Ml\cos(\alpha) \\ M\sin(\alpha) & M & 0 \\ Ml\cos(\alpha) & 0 & Ml^2 \end{bmatrix} \begin{bmatrix} \dot{r} \\ \dot{l} \\ \dot{\alpha} \end{bmatrix}$$

$$U(q) = -Mgl\cos(\alpha) \tag{5.81}$$

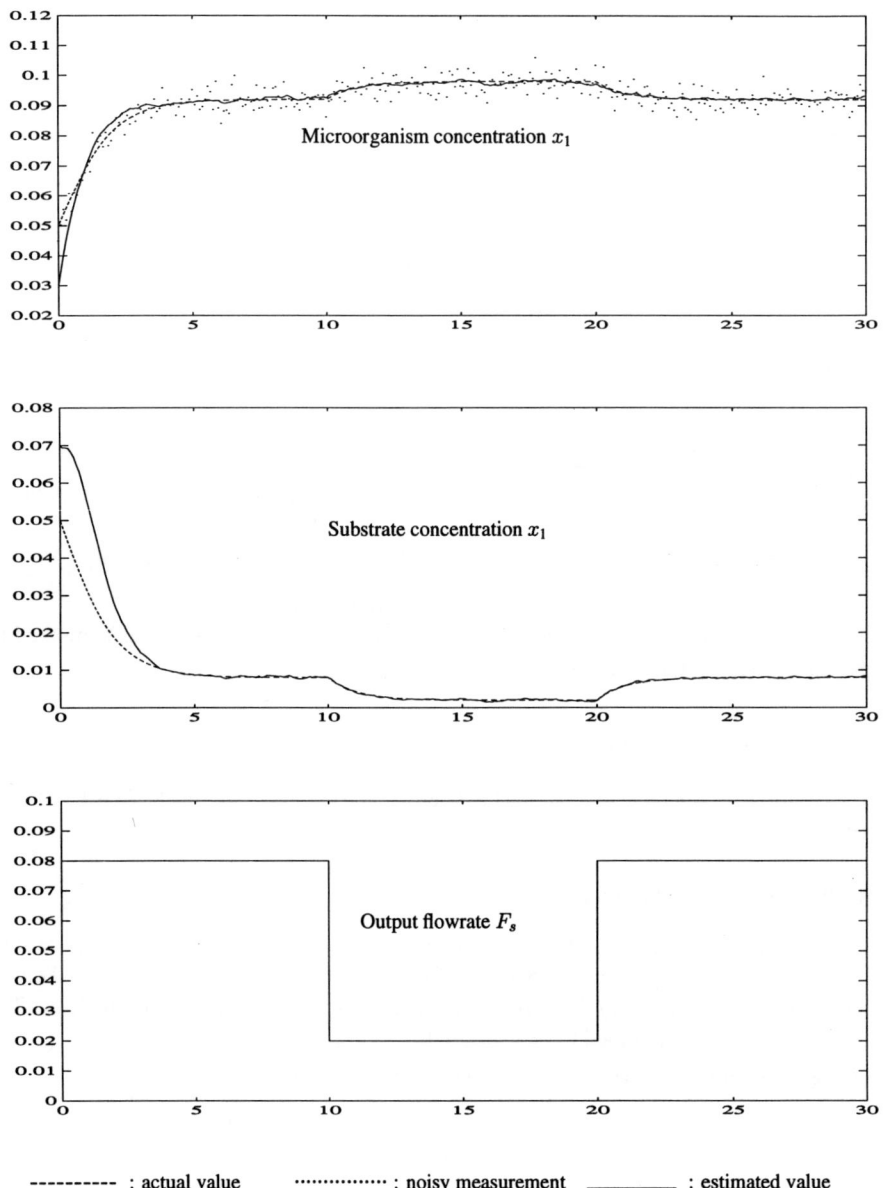

Figure 5.5: Bioreactor: simulation of the observer.

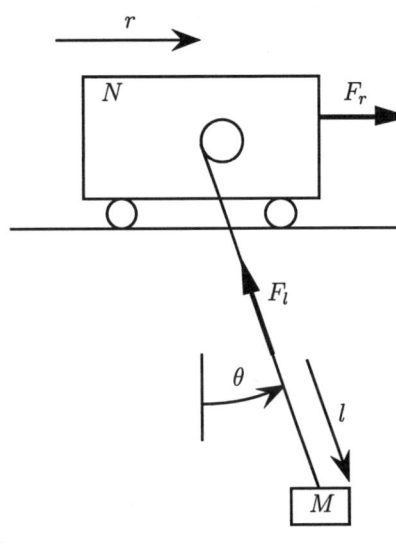

Notation:

M	:	Load
N	:	Carriage mass
r	:	Carriage position
l	:	Free length of the cable
α	:	Angle of the cable
F_r	:	Motion force on the carriage
F_l	:	Lifting force
L	:	Lagrangian
T	:	Kinetic energy
V	:	Potential energy
q	:	Generalized coordinates
F	:	Generalized force

Figure 5.6: Overhead crane.

The following equations are finally obtained:

$$
\frac{\mathrm{d}}{\mathrm{d}t}
\begin{bmatrix} r \\ \dot{r} \\ l \\ \dot{l} \\ \alpha \\ \dot{\alpha} \end{bmatrix}
=
\begin{bmatrix}
0 & 1 & & & \cdots & 0 \\
0 & 0 & & & & \vdots \\
& & 0 & 1 & & \\
& & 0 & 0 & & \\
\vdots & & & & 0 & 1 \\
0 & \cdots & & & 0 & 0
\end{bmatrix}
\begin{bmatrix} r \\ \dot{r} \\ l \\ \dot{l} \\ \alpha \\ \dot{\alpha} \end{bmatrix}
+
\begin{bmatrix}
0 \\
0 \\
0 \\
g\cos(\alpha) + l\dot{\alpha}^2 \\
0 \\
-\dfrac{g\sin(\alpha)}{l} - \dfrac{2\dot{l}\dot{\alpha}}{l}
\end{bmatrix}
$$

$$
+
\begin{bmatrix}
0 & 0 \\
\dfrac{1}{N} & \dfrac{-\sin(\alpha)}{N} \\
0 & 0 \\
\dfrac{-\sin(\alpha)}{N} & \dfrac{1}{M} + \dfrac{\sin^2(\alpha)}{N} \\
0 & 0 \\
\dfrac{-\cos(\alpha)}{Nl} & \dfrac{\sin(\alpha)\cos(\alpha)}{Nl}
\end{bmatrix}
\begin{bmatrix} F_r \\ F_l \end{bmatrix}
$$

$$
y =
\begin{bmatrix}
1 & 0 & & & \cdots & 0 \\
& & 1 & 0 & & \\
0 & \cdots & & 1 & 0
\end{bmatrix}
\begin{bmatrix} r \\ \dot{r} \\ l \\ \dot{l} \\ \alpha \\ \dot{\alpha} \end{bmatrix}
$$

(5.82)

5.8.2.2 Observer design

The structure of system (5.82) has exactly the form of (5.29). The structure assumptions (ii) and (iii) of theorem 8 are satisfied. In order to check, is sufficient to choose $\sigma = [121212]^T$.

Remark 12

This property is common to all the mechanical systems that are observed from their positions. This directly comes from the Lagrange formulation.

On the other hand, the assumption of a Lipschitzian function φ is not satisfied because of the terms $l\dot{\alpha}^2$, $l\dot{\alpha}$, $\dfrac{1}{l}$.

Nevertheless, it is possible to make the following boundedness assumptions: the control variables as well as the states (the positions and their first derivatives) are bounded – if not, the system would be subject to destruction.

Then it is sufficient to extend, outside a large enough compact domain Ω, the function φ of (5.82) by a Lipschitzian function. Theorem 8 thus applies for the modified system, and the corresponding observer can be designed.

Remark 13

The assumption made above does not guarantee that the estimated state will not have transient excursions out of Ω. It is sufficient that the real state remains in Ω in order to guarantee the observer convergence.

However, the observation-control coupled problem is not solved. The control is computed from the estimated state, and it is therefore impossible to guarantee that the bounddeness assumption is satisfied. The absence of a separation principle is here a concretization.

5.8.2.3 Results

The applicability conditions of the theorem are not strictly satisfied. Nevertheless if Ω is chosen large enough with respect to the accepted deviations for the state variables, the situation is as follows:

- Any acceptable control (for the physical system) is such that the convergence conditions are satisfied.
- In this case, the control "acceptability" is only tested by simulation.

The simulation results given in figure 5.7 have been obtained by applying a stabilizing control to the system (see Chorot, 1991).

The observer corresponding to theorem 8 has been built. For each subsystem, the loop gain K of the observer is chosen from the steady-state solution of a Lyapunov equation of the following form:

$$0 = -T^{-1}S - A^T S - SA + C^T C \tag{5.83}$$

whose explicit solution is:

$$S(T) = \begin{bmatrix} T & T^2 \\ T^2 & 2T^3 \end{bmatrix}, \quad S^{-1}(T) = \begin{bmatrix} 2T^{-1} & T^{-2} \\ T^{-2} & T^{-3} \end{bmatrix} \tag{5.84}$$

The responses of the figure 5.7 show a response time of the observer which is roughly half the response time of the control system. The behavior with respect to noisy measurements is good.

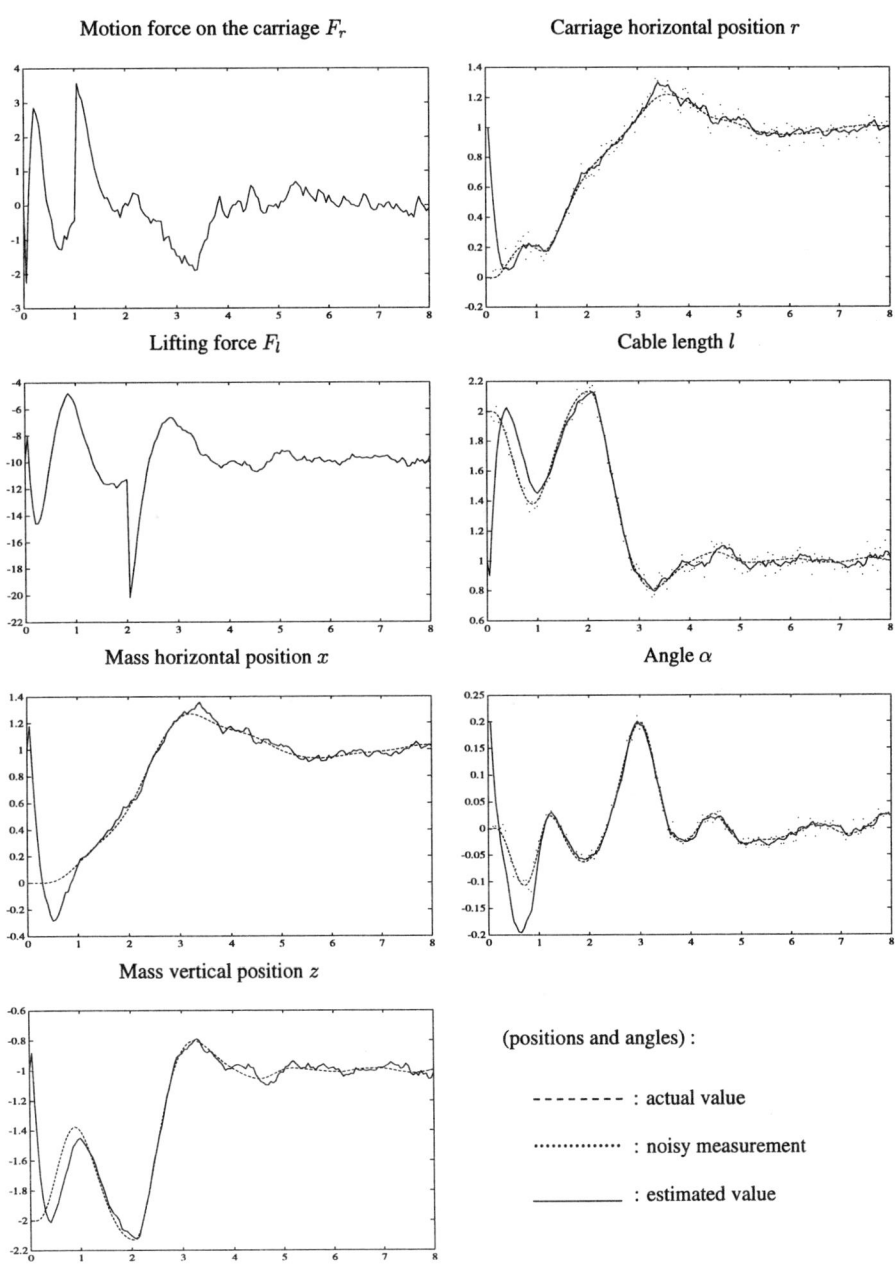

Figure 5.7: Overhead crane: simulation of the observer.

5.8.3   Distillation column

The example presented in this section[1] has been carried out by F. Deza at the Grand-Couronne Shell Research Center (Deza, 1991).

5.8.3.1   Process and model

The process which is studied is a binary distillation column. The scheme and the notations are given in figure 5.8.

The inputs are F, x_F (disturbances), L, V, (control variables). The measured outputs are w_1 and x_n. The input x_F is not measured.

Under Lewis assumptions, this system is described as follows (mass balances):

$$
\begin{cases}
\dot{x}_1 &= \dfrac{1}{A_1}V(w(x_2) - x_1) \\[2mm]
&\vdots \\[2mm]
\dot{x}_j &= \dfrac{1}{A_j}(L(x_{j-1} - x_j) + V(w(x_{j+1}) - w(x_j))) \\[2mm]
&\vdots \\[2mm]
\dot{x}_f &= \dfrac{1}{A_f}(L(x_{f-1} - x_f) + V(w(x_{f+1}) - w(x_f)) + F(x_F - x_f)) \\[2mm]
&\vdots \\[2mm]
\dot{x}_j &= \dfrac{1}{A_j}((L + F)(x_{j-1} - x_j) + V(w(x_{j+1}) - w(x_j))) \\[2mm]
&\vdots \\[2mm]
\dot{x}_n &= \dfrac{1}{A_n}((L + F)(x_{n-1} - x_n) + V(x_n - w(x_n))) \\[2mm]
\dot{x}_F &= 0 \\[2mm]
y_1 &= x_1 \\[2mm]
y_2 &= x_n
\end{cases}
\tag{5.85}
$$

where w represents the constant pressure liquid-vapor equilibrium law. It is a monotonous C^∞ diffeomorphism of $[0, 1]$ on itself, with $w(0) = 0$, $w(1) = 1$, $w(x) \geq x$, (non azeotropic equilibrium).

In this model, the unmeasured variable x_F is considered as a state variable and is given the simplest possible generating model – a constant prediction.

The admissible input domain $\mathcal{U}$ is characterized by the following inequalities:

$$
\begin{aligned}
&0 < L < L_{\max}, \quad 0 < V < V_{\max}, \quad 0 < F < F_{\max}, \\
&V - L > 0, \quad F + L - V > 0, \\
&0 < x_F < 1.
\end{aligned}
\tag{5.86}
$$

---

[1]We thank E. Busvelle, who provided the final simulations shown here

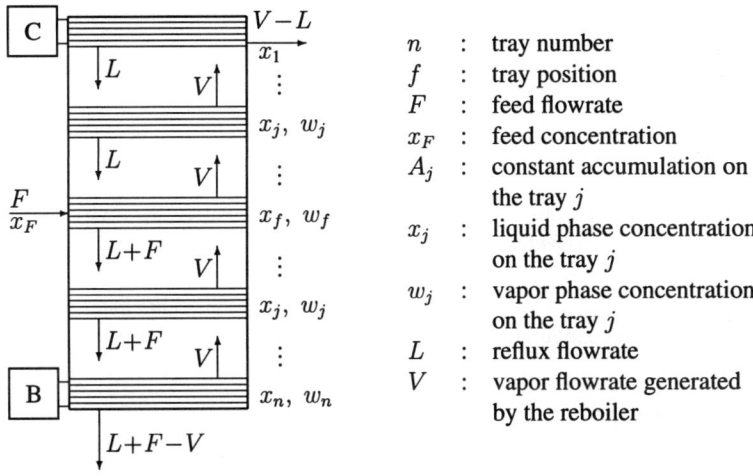

Figure 5.8: Distillation column.

The system is characterized by the following properties:

- It is affine with respect to the inputs L, V, F.

- The cube $[0, 1]^n$ is invariant under (5.85) for every admissible input, as well as its interior $\mathcal{X}$.

- Each constant input, corresponds to exactly one equilibrium state in $\mathcal{X}$, and this equilibrium is globally asymptotically stable on $\mathcal{X}$.

- Near each steady state, the linearized model is controlable, and each equilibrium state is accessible from any other state.

5.8.3.2 Observability

The following relationships are derived from the system equations:

$$
w(x_2) = x_1 + \frac{A_1 \dot{x}_1}{V}
$$

$$
\vdots
$$

$$
w(x_{j+1}) = w(x_j) + \frac{A_j \dot{x}_j - L(x_{j-1} - x_j)}{V}, \quad j = 2, \ldots, f-1 \tag{5.87}
$$

$$
\vdots
$$

$$
x_F = x_f + \frac{A_f \dot{x}_f - L(x_{f-1} - x_f) - V(w(x_{f+1}) - w(x_f))}{F}
$$

$$x_{n-1} = x_n + \frac{A_n \dot{x}_n - V(x_n - w(x_n))}{F + L}$$

$$\vdots$$

$$x_{j-1} = x_j + \frac{A_j \dot{x}_j - V(w(x_{j+1}) - w(x_j))}{F + L} \quad j = n-1, \ldots, f+2 \tag{5.88}$$

$$\vdots$$

The function w is a diffeomorphism. The inputs F, L, V are positive. Knowing these inputs and the outputs x_1, x_n, as well as their derivatives, the state variables $x_2, \ldots, x_{n-1}, x_F$ can be formally reconstructed.

The system is therefore observable for any admissible input. On the other hand, the inputs $V = 0$ and $F + L = 0$ are singular.

5.8.3.3 Observer design

We shall now set up coordinate transformations in order to put the system in a form of type (5.41) so as to test the conditions of theorem 9.

The first transformation simply consists of a subscript permutation for the states of the lower section:

$$\begin{cases} z_f &= x_n \\ z_{f+1} &= x_{n-1} \\ &\vdots \\ z_n &= x_f \end{cases} \tag{5.89}$$

Setting $z_{n+1} = x_F$ leads to:

$$\begin{cases} \vdots \\ \dot{x}_{f-1} &= \frac{1}{A_{f-1}}(L(x_{f-2} - x_{f-1}) + V(w(z_n) - w(x_{f-1}))) \\ \dot{z}_f &= \frac{1}{A_f}((L+F)(z_{f+1} - z_f) + V(z_f - w(z_f))) \\ \vdots \\ \dot{z}_j &= \frac{1}{A_j}((L+F)(z_{j+1} - z_j) + V(w(z_{j-1}) - w(z_j))) \\ \vdots \\ \dot{z}_n &= \frac{1}{A_n}(L(x_{f-1} - z_n) + V(w(z_{n-1}) - w(z_n)) + F(z_{n+1} - z_n)) \\ \dot{z}_{n+1} &= 0 \\ y_1 &= x_1 \\ y_2 &= z_f \end{cases} \tag{5.90}$$

which gives, for the lower section of the column:

$$
\begin{cases}
\dot{z}_f & = \dfrac{1}{A_f}(L+F)z_{f+1} + \varphi_f(L, V, F, z_f) \\[2mm]
& \vdots \\[2mm]
\dot{z}_j & = \dfrac{1}{A_j}(L+F)z_{j+1} + \varphi_j(L, V, F, z_f, \ldots, z_j) \\[2mm]
& \vdots \\[2mm]
\dot{z}_n & = \dfrac{1}{A_n}F z_{n+1} + \varphi_n(L, V, F, z_f, \ldots, z_n, x_{f-1}) \\[2mm]
\dot{z}_{n+1} & = 0
\end{cases}
\tag{5.91}
$$

The upper section contains terms of the form $w(x_j)$. Consider the following nonlinear coordinate change Φ:

$$
\begin{cases}
z_1 & = x_1 \\[1mm]
z_2 & = w(x_2) & \Rightarrow \dot{z}_1 & = \dfrac{1}{A_1}V(z_2 - z_1) \\[2mm]
& & & = \dfrac{1}{A_1}V z_2 + \varphi_1(V, z_1) \\[2mm]
z_3 & = w'(x_2)w(x_3) & \Rightarrow \dot{z}_2 & = w'(x_2)\dot{x}_2 \\[2mm]
& & & = \dfrac{1}{A_2}V z_3 + \dfrac{w'(x_2)}{A_2}(L(x_1 - x_2) \\[2mm]
& & & \qquad - V w(x_2)) \\[2mm]
& & & = \dfrac{1}{A_2}V z_3 + \varphi_2(L, V, z_1, z_2) \\[2mm]
z_4 & = w'(x_2)w'(x_3)w(x_4) & \Rightarrow \dot{z}_3 & = \dfrac{1}{A_3}V z_4 + \varphi_3(L, V, z_1, z_2, z_3) \\[2mm]
& \vdots \\[2mm]
z_{f-1} & = w'(x_2)\ldots w'(x_{f-2})w(x_{f-1}) & \Rightarrow \dot{z}_{f-2} & = \dfrac{1}{A_{f-2}}V z_{f-1} \\[2mm]
& & & \quad + \varphi_{f-2}(L, V, z_1, \ldots, z_{f-2}) \\[2mm]
& & \dot{z}_{f-1} & = \varphi_{f-1}(L, V, z_1, \ldots, z_{f-1}, z_n)
\end{cases}
\tag{5.92}
$$

where $w'(x)$ means $\dfrac{\partial w}{\partial x}(x)$.

Line n can be written as:

$$
\dot{z}_n = \frac{1}{A_n}F z_{n+1} + \varphi_n(L, V, F, z_1, \ldots, z_n)
\tag{5.93}
$$

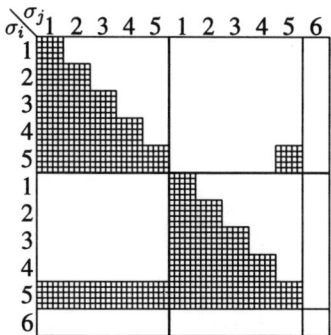

Figure 5.9: Structure of the Jacobian matrix.

These new coordinates lead, for the system (5.85), to the following equations:

$$
\begin{cases}
\dot{z}_1 &= \dfrac{1}{A_1}V z_2 + \varphi_1(V, z_1) \\[2mm]
&\vdots \\[2mm]
\dot{z}_j &= \dfrac{1}{A_j}V z_{j+1} + \varphi_j(L, V, z_1, \dots, z_j) \\[2mm]
&\vdots \\[2mm]
\dot{z}_{f-1} &= \varphi_{f-1}(L, V, z_1, \dots, z_{f-1}, z_n) \\[2mm]
\dot{z}_f &= \dfrac{1}{A_f}(L+F)z_{f+1} + \varphi_f(L, V, F, z_f) \\[2mm]
&\vdots \\[2mm]
\dot{z}_j &= \dfrac{1}{A_j}(L+F)z_{j+1} + \varphi_j(L, V, F, z_f, \dots, z_j) \\[2mm]
&\vdots \\[2mm]
\dot{z}_n &= \dfrac{1}{A_n}F z_{n+1} + \varphi_n(L, V, F, z_1, \dots, z_n) \\[2mm]
\dot{z}_{n+1} &= 0
\end{cases}
\tag{5.94}
$$

Figure 5.9 shows the structure of the Jacobian matrix of φ (with $n = 10$, $f = 6$).

Associated with the choice of indexes σ_i of figure 5.9, this structure clearly satisfies conditions (ii) and (iii) of theorem 9.

Remark 14

- *The coordinate transformation Φ is not global. As a matter of fact, the function w (liquid-vapor equilibrium) is only defined on the interval $[0, 1]$. Nevertheless, it can be extended to $\mathbb{R}$ by a C^∞ Lipschitzian function $\tilde{w}$.*

- *The coordinate change Φ defined with $\tilde{w}$ is global and the function φ is Lipschitzian. Moreover, $A(L, V, F)$ is bounded for bounded L, V, F, and C is constant. Condition (i) of the theorem is fulfilled.*

- *Finally, the lower bounds on $F + L$ and on V guarantee that the input is locally regular, which is enough to satisfy condition (iv) of theorem 9.*

5.8.3.4   Simulation test

The simulation was carried out for a water-methanol column whose characteristics are
given in table 5.1.

$\alpha = 1.65$	Tray number	27
$x_1 = 0.99$	Feed tray	14
$x_n = 0.01$	Condensor holdup	5.5
$F = 1.65$	Tray accumulation	0.55
$L = 4.69$	Reboiler accumulation	5.5
$V = 5.94$	x_F	0.75

Table 5.1: Water-methanol column - operating point.

The observer given by the theorem has been tested by simulation. As pointed out in section 5.5, this is an extended Kalman observer to which the implementation (coordinate choice, weighting matrices) gives special properties. Figure 5.10 shows the response of the proposed observer, compared to the natural response of the system and to a classical Kalman observer response.

5.9   Conclusions

In this chapter, we first pointed out a difficulty which is inherent in nonlinear systems observation: contrary to the linear case, the observability depends on the input and, in general, there exist singular inputs which lead to a lack of observability. This property is fundamental in the observers design. As a matter of fact, we must guarantee both that the studied system does not involve any singular input (uniformly observable systems), and that the inputs for which an observer works are specified.

Then we showed that Kalman type observers can be designed for state affine systems excited by inputs that are rich enough: the regularly persistent inputs.

Extensions then were proposed in two directions. The application of high gain type methods to strongly observable systems, as well as to some non-uniformly observable systems excited by inputs having a local regularity property, was presented. The use of immersion and output injection techniques for transforming a given system into one of the preceding forms was briefly exposed.

Some important problems still remain open, such as the definition of an observability canonical form for the uniformly observable multi-output systems, the separation principle problem recalled in this chapter, or the difficulty encountered in the calculation of coordinate transformations or immersions necessary for the synthesis.

At the end of this overview, we nevertheless know how to design an observer for systems which are miscellaneous and significant enough, of which some examples were described. From a practical point of view, we can apply one of the preceding methods, with caution, even when one of these conditions is not strictly satisfied. This was precisely the case in the overhead crane example.

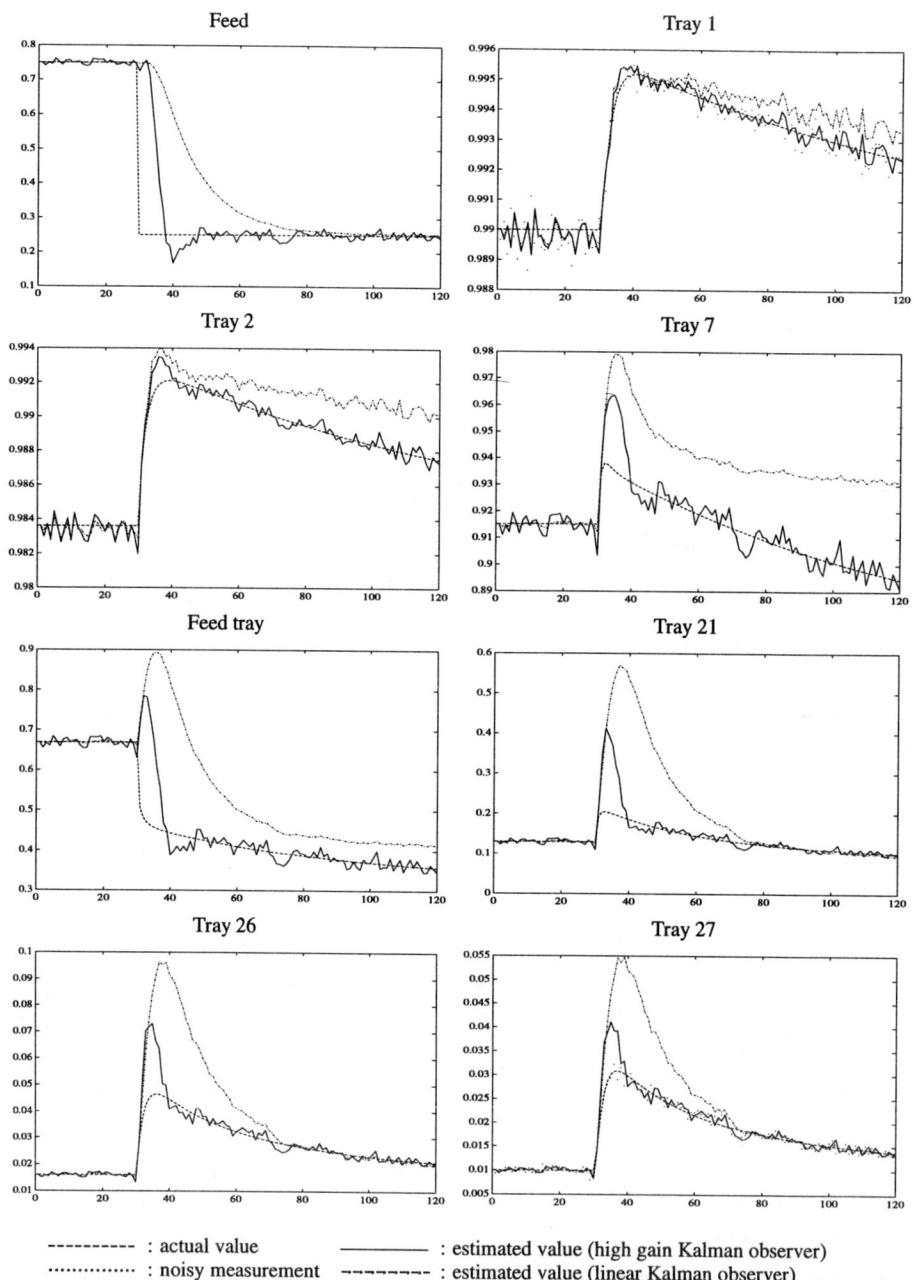

Figure 5.10: Distillation column: simulation of the observer.

5.10 Bibliography

[1] ANDERSON B.D. AND J. MOORE (1971). *Linear optimal control*, Prentice Hall Network series.

[2] BARBOT J.P. (1989). *Méthodes de calcul appliquées aux systèmes non-linéaires sous échantillonnage*, Doctorate thesis, Paris Sud University.

[3] BASTIN G. AND D. DOCHAIN (1990). *On line estimation and adaptive control of bioreactors*, Elsevier.

[4] BESTLE D. AND M. ZEITZ (1981). Canonical form design for nonlinear observers with linearizable error dynamics, *Internat. J. of Control*, 23, 419–431.

[5] BIRK J. AND M. ZEITZ (1988). Extended Luenberger observer for nonlinear multivariable systems, *Internat J. of Control*, 47 (6), 823–1836.

[6] BORNARD G. AND H. HAMMOURI (1991). *A high gain observer for a class of uniformly observable systems*, 30th IEEE CDC, Brighton.

[7] BORNARD G. AND H. HAMMOURI (1991). *An observer for a class of nonlinear systems under locally regular inputs*, internal note, LAG, France.

[8] BORNARD G., N. COUENNE AND F. CELLE (1988). *Regularly persistent observers for bilinear systems*, Proceedings of the 29th International Conference "New trends in nonlinear system theory", Nantes, Vol. 122, Springer Verlag.

[9] BOSSANE D., D. RAKOTOPARA AND J.P. GAUTHIER, *Local and global immersion into linear systems up to output injection*, 29th IEEE CDC, Honolulu, Hawai, 2000–2005.

[10] BRUNI C., G. DI PILLO AND G. KOCH (1971). On the mathematical models for bilinear systems, *Ricerche Automat.*, 2, 11–26.

[11] CELLE F., J.P. GAUTHIER, D. KASAKOS AND G. SALLET (1989). Synthesis of nonlinear observers: a harmonic analysis approach, *Math. System theory*, 22, 201–322.

[12] CELLE F., J. P. GAUTHIER AND D. KAZAKOS (1986), Orthogonal representations of nonlinear systems and input-output maps, *Sytems Control Lett.*, 7, 365–372.

[13] CHOROT T. (1991). *Modélisation, commande et observation des systèmes mécaniques hamiltoniens. Application à un pont roulant*, Doctorate thesis, Lyon 1 University, France.

[14] CHOROT T. AND J. DE LEON (1991). *Nonlinear stabilizing control law, with a dynamical high gain observer. Application to mechanical processes*, 30th IEEE - CDC, Brighton.

[15] CONTOIS D.(1959). Kinetics of the bacterial growth rate of continuous cultures, *Journal of Genetic Macrobiology*, 21, 40–50.

[16] COUENNE N. (1988), *Synthèse d'observateurs de systèmes affines en l'état*, Doctorate thesis, National Polytechnical Institute of Grenoble, France.

[17] CROUCH P. AND BYRNES C.(1988), Local accessibility, local reachability and representations of compact groups, *Math. Systems Theory*, 19, 43–65.

[18] DAVID R., J.M. DION AND I. D. LANDAU (1988). *Résultats et perspectives en Automatique*, Hermès, Paris.

[19] DEZA F. (1991), *Contribution à la synthèse d'observateurs exponentiels. Application à un procédé industriel: les colonnes à distiller*, Doctorate thesis, INSA de Rouen, France.

[20] DEZA F. AND J.P. GAUTHIER (1991). *Nonlinear observer for distillation columns*, 30th IEEE CDC, Brighton.

[21] DEZA F., E. BUSVELLE AND J.P. GAUTHIER (1992), Exponentially converging observers and internal stability of the dynamic output feedback for distillation columns, *Chemical eng. science*, vol. 47, 15–16, pp. 3935–3941.

[22] FLIESS M (1987). *Quelques remarques sur les observateurs non-linéaires*, Proceedings of the GRETSI, Nice, France.

[23] DIOP S. AND M. FLIESS (1991). *On nonlinear observability*, Proc. 1st European Control Conf., Grenoble, France, 152–157.

[24] FLIESS M. AND I. KUPKA (1983), A finiteness criterion for nonlinear input-output differential systems, *SIAM J. Control Optim.*, 21, 721–728.

[25] FUNAHASHI Y. (1979). Stable state estimator for bilinear systems, *Internat. J. of Control*. 29, 181–188.

[26] GAUTHIER J.P. AND G. BORNARD (1981). Observability for any u(t) of a class of bilinear systems, *IEEE Trans. Automat. Control*, AC 26, 922–926.

[27] GAUTHIER J.P. AND D. KASAKOS (1987). *Observabilité et observateurs de systèmes non- linéaires*, RAIRO APII, 21.

[28] GAUTHIER J.P. AND F. CELLE (1987). *Theory of dynamic observers for a class of nonlinear systems*, MTNS, Phoenix.

[29] GAUTHIER J.P., H. HAMMOURI AND I. KUPKA (1991). *Observers for nonlinear systems*, 30th IEEE CDC, Brighton.

[30] GAUTHIER J.P., H. HAMMOURI AND S. OTHMAN (1992). A simple observer for nonlinear systems. Application to bioreactors, *IEEE Trans. Auto. Control*, Vol. 36, n. 6.

[31] GRASSELI O. AND A. ISIDORI (1977). *Deterministic reconstruction and reachability of bilinear control processes*, Proc. JACC, San Francisco.

[32] GRASSELI O. AND A. ISIDORI (1981). An existence theorem for observers of bilinear systems, *IEEE Trans. Automat. Control*, AC 26, 1299–1301.

[33] HAMMOURI H., F. CELLE AND J. DE LEON (1991). *Cascade observers for some multi output nonlinear systems*, European Control Conf. HERMÈS, Grenoble.

[34] HAMMOURI H. AND J. DE LEON (1990). *On systems equivalence and observer synthesis*, New trends in Systems theory, Genoa.

[35] HAMMOURI H. AND J. DE LEON (1990). *Observer synthesis for state affine systems*, 29th IEEE CDC, Honolulu, Hawaii.

[36] HAMMOURI H. AND J.P. GAUTHIER (1988). Bilinearization up to output injection, *Systems and Control Letters*, 11, 139–149.

[37] HAMMOURI H. AND J.P. GAUTHIER (1989). *The time varying linearization up to output injection*, 28th IEEE CDC, Tampa, Florida.

[38] HAMMOURI H. AND J.P. GAUTHIER (1992). Global time varying linearization up to output injection, to be published, *SIAM J. Control Optim.*, 30, 6, 1295–1310.

[39] HARA S. AND K. FURUTA (1976). Minimal order state observers for bilinear systems, *Internat. J. of Control*, 24, 705–718.

[40] HERMANN R. AND A.J. KRENER (1977). Nonlinear controllability and observability, *IEEE Trans. Aut. Control.*, 22 (5), 728-740.

[41] KAILATH T. (1980). *Linear systems*, Prentice-Hall.

[42] KALMAN R.E. (1960). A new approach to linear filtering and prediction problems, *Transactions of ASME, Journal of Basic Engineering, D.*, 82, 35–45.

[43] KALMAN R.E. AND R. BUCY (1960). New results in linear filtering and prediction theory, *J. of Basic Engineering D.*, 82, 35–40.

[44] KALMAN R.E., P. FALB AND M. ARBIB (1969). *Topics in mathematical systems theory*, McGraw Hill.

[45] KASAKOS D (1987). *Sur l'Observabilité et les observateurs de Systèmes Non- Linéaires*, Thesis, National Polytechnical Institute of Grenoble, France.

[46] KOU S.R., D.L. ELLIOT AND T.J. TARN (1975). Exponential observers for nonlinear dynamic systems, *Inform. and Control*, 29, 204–216.

[47] KRENER A.J. AND A. ISIDORI (1983). Linearization by output injection and nonlinear observers, *Systems and Control Letters*, 3, 47–52.

[48] KRENER A.J. AND W. RESPONDEK (1985). Nonlinear observers with linear error dynamics, *SIAM J. Control and Optim.*, 23,197–216.

[49] KWAKERNAAK H. AND R. SIVAN (1972). *Linear Optimal Control Systems*, Wiley.

[50] LAGGOUNE N. AND G. GILLES (1984). *Observateur discret d'une classe de systèmes linéaires continus à paramètres variables*, Workshop on nonlinear system theory, Grenoble, France.

[51] LEVINE J. AND R. MARINO (1986). Nonlinear system immersion, observers and finite dimensional filters, *Systems and Control Letters*, 7, 137–142.

[52] LUENBERGER D.G. (1966). Observers for multivariable systems, *IEEE trans. on Aut. Control*, AC-11, 190–197.

[53] MONACO S. AND D. NORMAND-CYROT (1985). *On the sampling of a linear analytic control system*, 24th Conf. on Decision and Control, Fort Lauderdale, 1457–1462.

[54] NIJMEIJER H. (1982). Observability of a class of nonlinear systems: a geometric approach, *Systems and Control Letters*.

[55] O'REILLY J. (1983). *Observers for linear systems*, Academic Press.

[56] OTHMAN S. (1992). *Sur les observateurs des systèmes non linéaires*, Doctorate thesis, Lyon 1 University.

[57] SONTAG E. (1979). On the observability of polynomial systems, *SIAM J. of Control and Optim.*, 17.

[58] SUSSMAN H.J. (1979). Single input observability of continuous time systems, *Math. Systems Theory*, 12, 371–393.

[59] TORNAMBE A. (1989). *Use of asymptotic observers having high gains in the state and parameter estimation*, 28th IEEE CDC, Tampa.

[60] TSINIAS J. (1990). Further results on the observer design problem, *Syst. and Control. Letters*, 14.

[61] VAN DER SCHAFT A.J. (1985). On nonlinear observers, *IEEE Trans. on Auto. Control*, 30.

[62] WILLIAMSON D. (1977). Observability of bilinear systems with application to biological control, *Automatica*, 13.

[63] XIAO HUA XIA AND WEI BIN GAO (1989). Nonlinear observer design by observer error linearization, *SIAM J. of Control and Optim.*, 27, 1.

MASSON, Éditeur
120, bd Saint-Germain
75280 Paris Cedex 06
Dépôt légal : avril 1995

CORLET, Imprimeur, S.A.
14110 Condé-sur-Noireau
N° d'Imprimeur : 9353
Dépôt légal : mars 1995